MATHEMATICS
COURSE BOOK

Mike Smith

First published in 2019 by:
Bright Red Publishing Ltd
1 Torphichen Street
Edinburgh
EH3 8HX

A CIP record for this book is available from the British Library.

ISBN 978-1-84948-317-9

With thanks to:

Cover design, series book design, layout and setting by Caleb Rutherford – e i d e t i c.

Julie Bond (copy-edit). With special thanks to Alice Piotrowska for her illustration work.

Acknowledgements

Every effort has been made to seek all copyright-holders. If any have been overlooked, then Bright Red Publishing will be delighted to make the necessary arrangements.

Permission has been sought from all relevant copyright holders and Bright Red Publishing are grateful for the use of the following:

Mr Aesthetics/Shutterstock.com (p 5); Lotus_studio/Number1411/Shutterstock.com (pp 6–7); Ljupco Smokovski/Shutterstock.com (p 10); mchudo/iStock.com (p 10); urfin/Shutterstock.com (p 12); oksana2010/Shutterstock.com (p 17); Photo by Mircea Iancu from Pexels (p 18); Phil Talbot/Alamy Stock Photo (p 22); 4zevar/Shutterstock.com (p 23); iQoncept/Shutterstock.com (p 24); John A Cameron/Shutterstock (p 25); freeimages.com (p 26); nikamata/iStock.com (p 29); Dennis Jarvis (CC BY-SA 2.0)[2] (p 32); cornfield/Shutterstock.com (p 33); Skype is a trade mark or other intellectual property of the Microsoft group of companies and Bright Red Publishing is not affiliated, sponsored, authorized, or otherwise associated with or by the Microsoft group of companies (p 33); Nerthuz/Shutterstock.com (p 34); Oleksandra Naumenko/Shutterstock.com (p 34); sommersby/iStock.com (p 37); Katy Tzaralunga (CC BY 2.0)[1] (p 38); CaseyMartin/Shutterstock.com (p 44); swissmediavision/iStock (p 44); bigjom jom/Shutterstock.com (p 48); Marco Iacobucci Epp/Shutterstock.com (p 49); Two images by Boca Tutor (CC BY-SA 3.0)[3] (p 50); Perseus1984 (CC BY-SA 4.0)[4] (p 50); Daniel Stockman (CC BY-SA 2.0)[2] (p 57); Javier Brosch/Shutterstock.com (p 59); Jess Alford/Photodisc (p 60); Chamille White/Shutterstock.com (p 61); Nicole Gordine (CC BY 3.0)[5] (p 63); Nejron Photo/Shutterstock.com (p 64); DJM-photo/iStock.com (p 66); Antonio Gravante/Shutterstock.com (p 70); Ruslan Shevchenko/Shutterstock.com (p 70); Lepas/Shutterstock.com (p 71); studioworkstock/Shutterstock.com (p 71); urbanbuzz/Shutterstock.com (p 73); P.Burghardt/Shutterstock.com (p 73); Oleksiy Mark/Shutterstock.com (p 74); SpeedKingz/Shutterstock.com (p 76); Halfpoint/Shutterstock.com (p 77); iQoncept/Shutterstock.com (p 79); Dreamstime.com (p 80); lcswart/Shutterstock.com (p 81); Luca Galuzzi (CC BY-SA 2.5)[6] (p 82); Charles Lanteigne (CC BY-SA 3.0)[3] (p 82); saiko3p/Shutterstock.com (p 83); tansaisuketti (CC BY-SA 3.0)[3] (p 84); freeimages.com (p 86); Iakov Filimonov/Shutterstock.com (p 88); Segway is the registered trademark of Segway Inc. (p 88); Tommy Alven/Shutterstock.com (p 89); aldorado/Shutterstock.com (p 94); Pete Stewart (CC BY-SA 2.0)[2] (p 95); travelview/Shutterstock.com (p 96); allanswart/iStock.com (p 102); Critchleyhope (CC BY-SA 4.0)[4] (p 104); Jpquidores (CC BY-SA 3.0)[3] (p 107); Lowes (CC BY-SA 4.0)[4] (p 109); Dvest/Dreamstime.com (p 112); KobchaiMa/Shutterstock.com (p 119); Klaus Barner (CC BY-SA 3.0)[3] (p 121); Clément Bardot (CC BY-SA 3.0)[3] (p 122); Derek Ramsay (CC BY-SA 2.5)[6] (p 122); Jason Hollinger (CC BY 2.0)[1] (p 122); Alvesgaspar (CC BY-SA 3.0)[3] (p 122); ESA/Hubble (CC BY-SA 3.0)[3] (p 123); artskvortsova/Shutterstock.com (p 123); Lars Poyansky/Lorelyn Medina/Shutterstock.com (p 123); xamnesiacx/Shutterstock.com (p 125); Kovalov Anatolii/Shutterstock.com (p 128); Petr Vaclavek/Shutterstock.com (p 130); Andis Rea/Shutterstock.com (p 132); Natursports/Shutterstock.com (p 133); iStock.com (p 134); Emily Li/Shutterstock.com (p 139); Zheng Bin – Dreamstime.com (p 142); Africa Studio/Shutterstock.com (p 145); skvoor/Shutterstock.com (p 151); Denis Kovin/Shutterstock.com (p 154); Three images by Oleksandr Kostiuchenko/Shutterstock.com (p 155); Nikandphoto/Shutterstock.com (p 155); irin-k/Shutterstock.com (p 155); Jojje/Shutterstock.com (p 159); Dionisvera/Shutterstock.com (p 160); PowerUp/Shutterstock.com (p 173); Caleb Rutherford/RH Publications (p 182); Cid Jacobo/Creative Commons (CC BY-SA 4.0)[4] (p 183); NordNordWest/Creative Commons (CC BY-SA 3.0 de)[7] (p 185); Leonid Andronov/Shutterstock.com (p 185); Airfix is a registered trademark and the property of Hornby Hobbies Ltd (p 186); studiocasper/iStock.com (p 186); Jacob Blount/Shutterstock.com (p 186); Lenscap Photography/Shutterstock.com (p 189); Popartic/Shutterstock.com (p 190); Ewelina Wachala/Shutterstock.com and Wavebreakmedia/Shutterstock.com (p 191); Doodledoo/Creative Commons (CC BY-SA 3.0)[3] (p 191); Caleb Rutherford/Mirexon/Shutterstock.com (p 193); DmitriyRazinkov/Shutterstock.com (p 202); Vladimirkarp/Shutterstock.com (p 203); Kedofoto/Shutterstock.com (p 203); Pascal Halder/Shutterstock.com (p 203); Ivan Smuk/Shutterstock.com (p 203); Alexey Kljatov/Shutterstock.com (p 203); EvrenKalinbacak/Shutterstock.com (p 209); cynoclub/Shutterstock.com (p 212); ILYA AKINSHIN/Shutterstock.com (p 218); Standret/Shutterstock.com (p 222); vesna cvorovic/Shutterstock.com (p 225); JaneHYork/Shutterstock.com (p 229); M. Unal Ozmen/Shutterstock.com (p 230); Antonio Guillem/Shutterstock.com (p 231); Vladvm/Shutterstock.com (p 233); Olga Popova/Shutterstock.com (p 235); Adisa/Shutterstock.com (p 236); chrisdorney/Shutterstock.com (p 236); urbanbuzz/Shutterstock.com (p 239); Rob Chandler/Creative Commons (CC BY 2.0)[1] (p 240); Image(s) licensed by Ingram Image (pp 1, 4, 8–11, 13–14, 16, 31, 33, 44–5, 60, 65, 68–70, 78, 85, 87, 89, 93, 95–9, 101–2, 105–7, 109–10, 127, 129, 136, 173, 178, 184, 193, 198, 217, 220–1, 226, 240–1; Caleb Rutherford (pp 4–5, 8, 13, 27, 43, 49, 55–6, 81, 155, 174, 235–6, 256).

Printed and bound by Replika Press Pvt. Ltd.

CONTENTS

INTRODUCTION

UNIT 1 – NUMBER, MONEY AND MEASURE

UNIT 2 – SHAPE, POSITION AND MOVEMENT

UNIT 3 – INFORMATION HANDLING

APPENDICES

INTRODUCTION

PATHWAY TO SUCCESS

Welcome to this BGE 3rd level textbook! There is a saying 'Practice makes Perfect'; this book says **'Practice makes Progress!'**

Treat mathematics as a 'verb'; it is a 'doing' word!

What can you expect to find in this book?

All the experiences and outcomes in 3rd level are covered and all of these are matched to ***Benchmarks***.

The book is made up of three core sections: *'Number, money and measure'*, *'Shape, position and movement'* and *'Information handling'*.

In each section you can find a wealth of worked examples, graduated Classroom Challenges, Investigations, Homework Helpers and Stretch Yourself questions to challenge your learning.

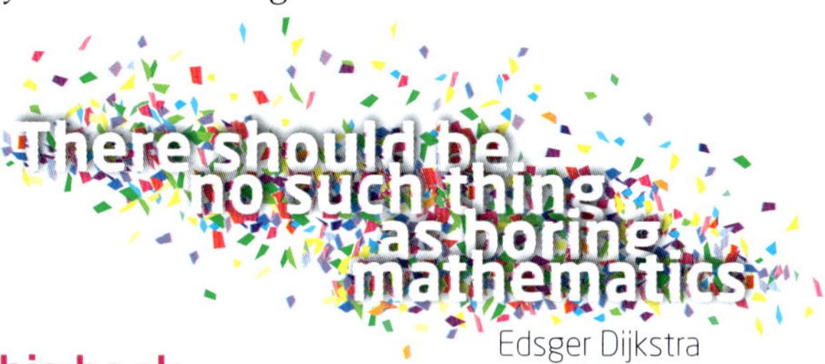

Features you will find in this book

Chapter title

This highlights the skill(s) you will cover in a chapter, including the experience and outcome.

What's coming up?

This explains the benchmark(s) linked with the topic and skills you are learning.

What you already know

This highlights what you have learned already – so you can see how you are making progress in the topic.

How does that work?

Worked examples which explain methods, strategies and techniques for answering questions and solving the problems in the Classroom Challenges.

Classroom challenge

Almost every page has at least one Classroom Challenge, a collection of bright and colourful exercises with plenty of examples to build up and consolidate your skills. These exercises help you to put together a 'toolkit' useful for working on related or more challenging problems

STRETCH YOURSELF

A question or investigation which encourages you to show you fully understand the skill you have been learning.

TEACHING NOTES

Templates, guides and notes can be downloaded for free from the Bright Red Publishing website (www.brightredpublishing.co.uk)

HOMEWORK HELPERS

This free bundle collates further material available, extra research you can do and, just to make sure your knowledge is fully secure, additional exercises you can do at home to ensure your new information or skill is as firmly 'locked in' as you think it is.

The practice you'll get in 'Classroom challenges' and 'Homework helpers' is especially designed to sharpen your understanding of the wording and types of questions you will be expected to answer in any assessment or progress check. And, even more importantly, to help your understanding of how to best answer them!

Finally, you can find useful tips, hints and important information highlighted in our **Don't Forget** pointers.

Hopefully this book will take you along your pathway to success, and remember:

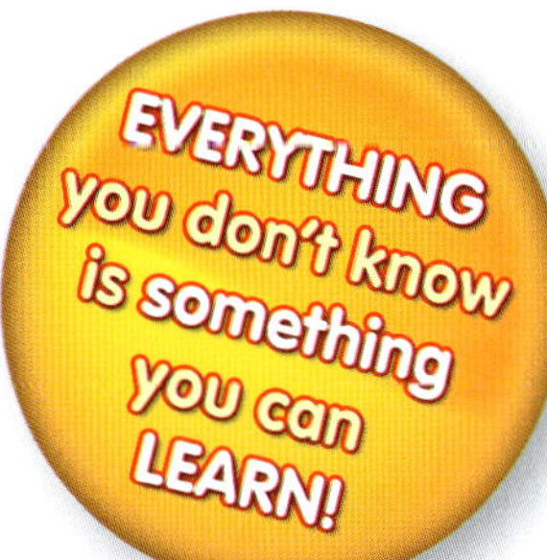

Good Luck with this 3rd Level Course – we look forward to seeing you at the 4th Level!

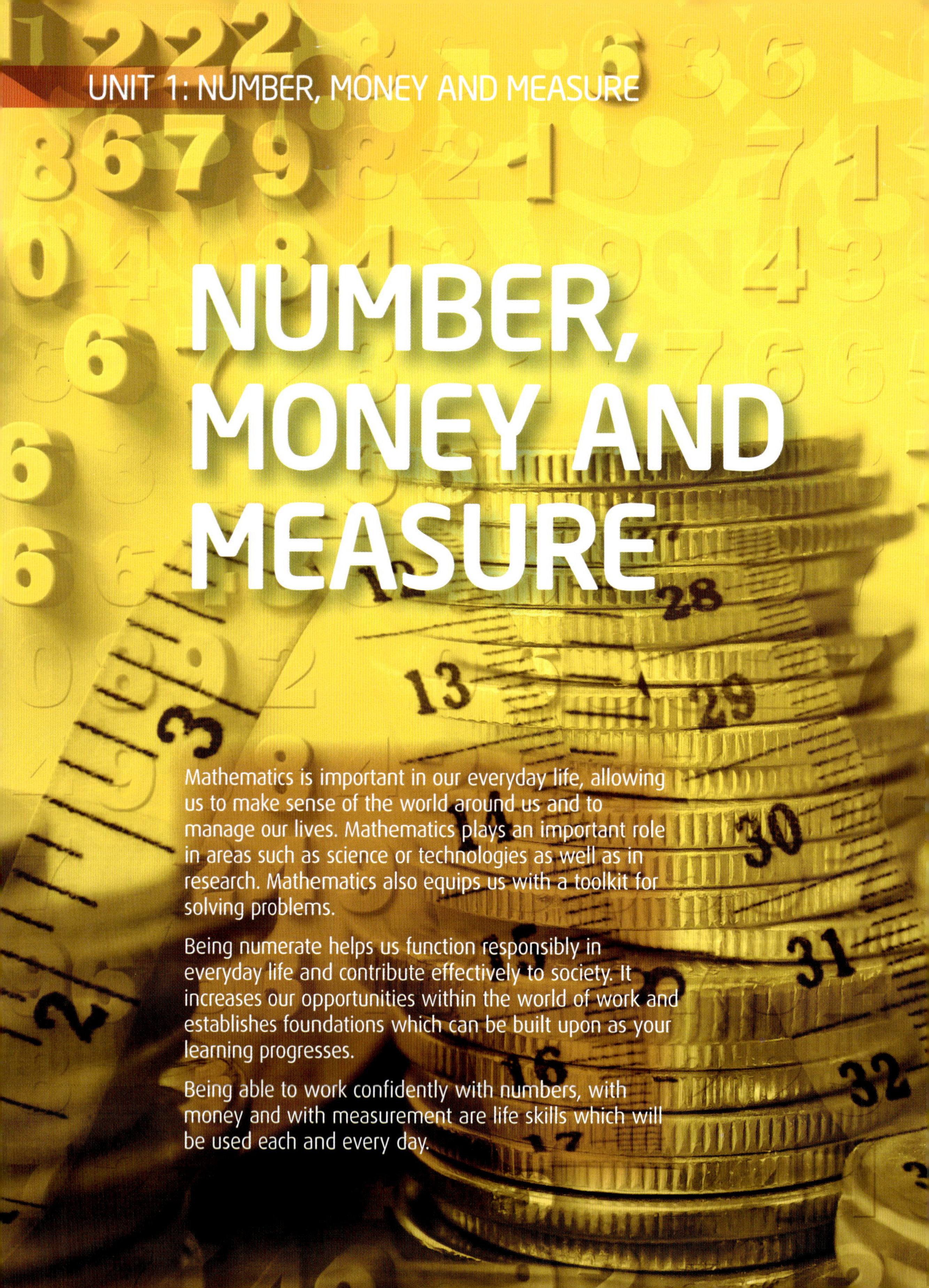

NUMBER, MONEY AND MEASURE

Mathematics is important in our everyday life, allowing us to make sense of the world around us and to manage our lives. Mathematics plays an important role in areas such as science or technologies as well as in research. Mathematics also equips us with a toolkit for solving problems.

Being numerate helps us function responsibly in everyday life and contribute effectively to society. It increases our opportunities within the world of work and establishes foundations which can be built upon as your learning progresses.

Being able to work confidently with numbers, with money and with measurement are life skills which will be used each and every day.

UNIT 1: NUMBER, MONEY AND MEASURE

NUMERACY SKILLS

ESTIMATION AND ROUNDING

I can round a number using an appropriate degree of accuracy, having taken into account the context of the problem. MNU 3-01a

What's coming up?

This Outcome and Experience will give you the opportunity to:

- round decimal fractions to three decimal places
- use rounding to routinely estimate the answers to calculations.

What you already know

You have already learned how to:

- ✔ round to nearest whole number
- ✔ round to nearest 10, 100, 1000, 10 000, 100 000
- ✔ round to 1 or 2 decimal places
- ✔ apply knowledge of rounding to give an estimate to a calculation appropriate to the context.

Rounding to three decimal places

In mathematics, we do not always have to be exact.

Sometimes we may wish to round a number.

For example, if you are working with money you would round your answer to two decimal places – as that would be the 'pence' columns.

How does that work?

4 friends share a restaurant bill of £128·71 equally between them.

How much would each pay?

SOLUTION

£128·71 ÷ 4

= £32·1775

This is too accurate as we cannot split money into such small units.

So we round £32·1775

We want 2 decimal places, so we look at the 3rd decimal place. Using '5 or more we round it up, less than 5 we leave it', we would round this to

£32·18

Therefore, if we want to round to 3 decimal places we would look at the 4th decimal place and use our '5 or more' rule.

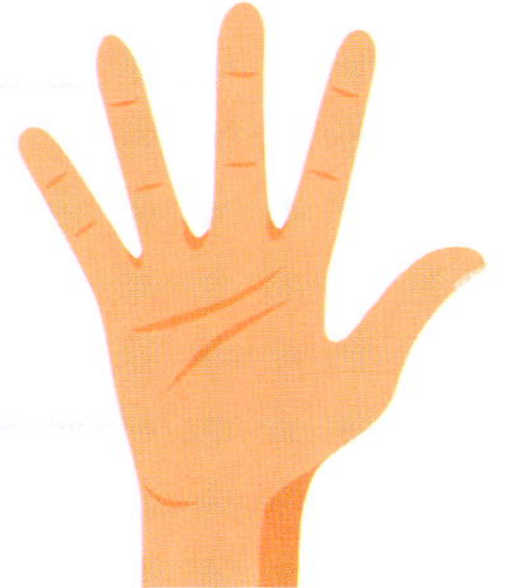

How does that work?

Round 146·78347 to 3 decimal places.

SOLUTION

146·78347

= 146·78347   4 is less than 5 so the previous digit stays as it is

= 146·783   to 3 decimal places (3dp)

Classroom challenge

Currency exchange rates are often given to 4 or more decimal places.

Look at the currency exchange rates in the table below and round each currency to 3 decimal places.

The first one has been done for you.

£1 GBP (Great Britain Pound) buys		
Currency	**Exchange rate**	**Written correct to 3 dp**
Euro €	1·08011	1·080

£1 GBP (Great Britain Pound) buys		
Currency	**Exchange rate**	**Written correct to 3 dp**
US dollar **$**	1·2937	
Indian rupee ₹	82·6011	
Australian dollar **A$**	1·6244	
Arab Emirates dirham د.إ	4·75300	
Bulgarian lev **лв**	2·11725	
Chinese yuan **¥**	8·55679	
Egyptian pound ج.م	22·8910	
Hong Kong dollar **HK$**	10·1189	
Moroccan dirham **MAD**	12·0645	

Using rounding to estimate an answer

When doing calculations, it is useful to have a rough idea of the expected 'size' of the answer. For example, when you are shopping in a supermarket, you should have a rough idea how much your shopping is going to come to; is it likely to be £50 or £100 or £200, so that when the total comes up on the till it should not be a surprise.

How does that work?

EXAMPLE

Mark buys 9 pieces of fencing at £28·47 each.
Roughly how much will this cost?

SOLUTION

Actual 'sum' is 9 × £28·47
This is about 10 × £30 which is £300
So Mark would expect his bill to be about £300
Compare this with the actual bill of £256·23

EXAMPLE

3612 kilograms of potatoes are put into 8 boxes. Approximately how many kilograms would go into each box?

SOLUTION

If you are confident in dividing by a single digit you may say:

Actual sum 3216 ÷ 8
is approximately 3200 ÷ 8 = 400 kg

Or you may be more confident dividing by 10 and so may say:

Actual sum 3216 ÷ 8
is approximately 3200 ÷ 10 = 320 kg
or 3000 ÷ 10 = 300 kg

In either method you can see your answer should be 'in the hundreds'.
Compare with actual answer: 3216 ÷ 8 = 402 kg

You are looking for a '**size**' of answer; is it in tens, hundreds or thousands?

Classroom challenge

1. By rounding each number, estimate the answer to these calculations. The first one has been done for you.

Calculation	Rounded numbers	Estimate answer	Actual answer
18 × 47·3 mm	20 × 50	1000 mm	851·4 mm
12 × £31·15			
104 × 9·8 g			
387 kg ÷ 19·17			
47·2 + 104·1 + 32·7			
12121 ÷ 8			
23·1 × 9·2 + 73·5			

2. There are 29 students in each of the 8 First Year classes in Allanmuir High School. Approximately how many students are there in First Year at Allanmuir High School?

3. 2416 tickets were sold for 8 performances of a play in a theatre. Approximately how many tickets were sold for each performance?

4. Mark bought 4 items from a shop. The items cost £5·82, £9·99, £0·32 and £4·12
To the nearest £1, how much would Mark expect to pay in total?

5. At a sports event, 5 competitors took part in the javelin. The distances they threw the javelin were recorded as 58·13 m, 61·07 m, 47·93 m, 51·65 m and 32·07 m
What was the approximate total distance thrown?

NUMBER AND NUMBER PROCESSES

Number problems in familiar contexts

I can use a variety of methods to solve number problems in familiar contexts, clearly communicating my processes and solutions. MNU 3-03a

What's coming up?

This Outcome and Experience will give you the opportunity to:

- recall quickly multiplication and division facts to the 10th multiplication table
- use multiplication and division facts to the 12th multiplication table.

What you already know

You have already learned how to:

- ✔ use multiplication and division facts to the 10th multiplication table
- ✔ multiply and divide whole numbers by multiples of 10, 100 and 1000
- ✔ multiply whole numbers by two-digit numbers.

Recalling my times tables

Being able to work quickly and accurately with 'numbers' is a very useful skill to have.

The ability to work both 'mentally' and using written methods will be of great benefit both in other subjects at school and in later life.

In this section we will practise our 'mental arithmetic' as well as solving problems using our number skills.

How does that work?

Quickly recalling number bonds or 'times tables' helps us to solve problems more quickly and more confidently. If you were asked 'what is 7×8?' what would you do?

Think of times tables as exactly that – a table with rows of plates on it.

So, 7 × 8 is a table with 7 rows each with 8 plates.

Getting this 'mental picture' will help.

We can see that 7 × 8 = 56.

There is more practice on the Bright Red website. Download our Homework Helpers and keep practising!

Classroom challenge

There are two multiplication squares in our Teaching Notes: one up to 10 × 10 and one up to 12 × 12.

1 Speed and accuracy test!
Try these sets of times tables questions.
Take a note of how long you took to do each set, and how accurate you are.

Set 1	Set 2	Set 3	Set 4	Set 5	Set 6
3 × 4	6 × 2	8 × 4	10 × 3	12 × 3	1 × 1
4 × 4	6 × 8	8 × 7	11 × 5	12 × 6	2 × 2
5 × 3	7 × 4	8 × 12	5 × 11	12 × 8	3 × 3
4 × 7	7 × 7	8 × 6	10 × 9	12 × 5	4 × 4
3 × 10	7 × 10	9 × 9	9 × 10	12 × 10	5 × 5
4 × 11	6 × 12	9 × 11	10 × 11	12 × 7	6 × 6
5 × 12	7 × 11	9 × 5	11 × 12	12 × 12	7 × 7
4 × 12	6 × 10	9 × 7	10 × 12	12 × 2	8 × 8
3 × 11	6 × 11	9 × 12	11 × 10	12 × 1	9 × 9
5 × 11	7 × 12	8 × 11	11 × 11	12 × 0	10 × 10

2 Speed and accuracy again!
This time there is a mixture of multiplying and dividing.
Look carefully.

Set 1	Set 2	Set 3
72 ÷ 9	9 × 4	144 ÷ 12
8 × 7	36 ÷ 9	10 × 10
44 ÷ 4	42 ÷ 7	88 ÷ 11
36 ÷ 4	6 × 7	48 ÷ 12
5 × 9	5 × 7	55 ÷ 5
10 × 6	35 ÷ 7	5 × 11
48 ÷ 8	6 × 11	121 ÷ 11

Working with numbers in familiar contexts

When solving problems, you should work through them in a logical manner.

How does that work?

Carol can type at a speed of 52 words per minute.

How many words could she type in 7 minutes?

- READ the question carefully.
- Select a STRATEGY (or how you are going to tackle the question).
- Decide on the OPERATION(S) to use. (Is it add, subtract, multiply, divide or a mixture?)
- Decide what the ANSWER is going to look like.

DON'T FORGET

You will have learned a number of strategies for solving problems. For example, set out a calculation, make a table, draw a diagram, simplify the problem, make a list, work backwards, look for a pattern.

SOLUTION

- Read the question at least twice; once quickly to get a 'feel' for it then again to look at the detail.
- Strategy: Where have I seen a question like this before? How did I do that one? Is it a 'calculation', a 'draw a diagram', a 'make a table' etc.
- Operations: The 'per' and the wording suggests we need to 'multiply up'.
- Answer: This is going to look like, *'answer' words in 7 minutes.*
- How your working would look:

Carol could type 364 words in 7 minutes.

How does that work?

A packet of sweets weighs 25 grams.

What would be the weight of 200 packets of sweets?

SOLUTION

It is a multiplying question.

You need to calculate 25×200
$= 25 \times 2 \times 100$
$= 50 \times 100$
$= 5000$

The weight would be 5000 grams.

Classroom challenge

Using the method above, solve these problems.

Remember to write your answer fully and not 'just a number'.

1 One litre of milk costs 95p.
How much will 5 litres of milk cost?

2 Jane's place of work is 15 km from her home.
How far will Jane travel if she makes 5 **return** trips to work?

DON'T FORGET

5 return trips is the same as 10 single trips

3 A recipe says that 1 kg of chocolate is needed to make 6 chocolate muffins.
How many kilograms of chocolate will be needed to make 24 chocolate muffins?

4 A kitchen worktop is 3000 mm in length.
A kitchen cupboard is 600 mm wide.
How many cupboards would the worktop cover?

DON'T FORGET

You may wish to review how you multiply and divide by multiples of 10, 100 and 1000.

5 Trish's car travels 42 km on 1 litre of petrol.
How far could she travel on a tank which has 30 litres of petrol in it?

6 EasyAir have 8 flights per day leaving Edinburgh Airport.
Each plane can carry up to 236 passengers.
How many passengers could fly out of Edinburgh, with EasyAir, in a day?

7 Akib has a collection of 132 CDs.
His CD storage rack can hold 12 CDs on each shelf.
How many shelves will Akib need to store all of his CDs?

8 The distance from the Earth to the Moon is about 400 000 kilometres.
How many kilometres would a Space shuttle cover if it flew to the Moon, and back, three times?

9 It costs £72 for a replica football kit.
How much would it cost for 11 such kits?

10 It costs £63 for one night in a hotel.
How much would it cost for 12 nights?

11 Lottery winning of £4500 is shared equally between 30 people.
How much did each person receive?

12 Lionel Messi is paid about $200 000 per week.
How much would he be paid over 52 weeks?

13 1375 ml of juice is shared between 11 glasses.
How many ml are in each glass?

14 Tony gets paid £240 per week.
How much would he get paid for 24 weeks?

15 A plank of wood is 142 cm long.
What would be the total length of 17 planks?

16 A box contains 42 drawing pins.
How many drawing pins would 36 boxes contain?

Using number facts in calculations

I can continue to recall number facts quickly and use them accurately when making calculations. MNU 3-03b

What's coming up?

This Outcome and Experience will give you the opportunity to:

- solve addition and subtraction problems working with whole numbers and decimal fractions to three decimal places
- solve multiplication and division problems working with whole numbers and decimal fractions to three decimal places.

What you already know

You have already learned how to:

- ✔ add and subtract whole numbers and decimal fractions to two decimal places, within the number range 0 to 1 000 000
- ✔ multiply and divide whole numbers by multiples of 10, 100 and 1000
- ✔ multiply and divide decimal fractions to two decimal places by 10, 100 and 1000
- ✔ multiply decimal fractions to two decimal places by a single digit
- ✔ divide whole numbers and decimal fractions to two decimal places, by a single digit, including answers expressed as decimal fractions, for example, 43 ÷ 5 = 8·6
- ✔ apply the correct order of operations in number calculations when solving multi-step problems.

Adding and subtracting whole numbers and decimals (up to 3 decimal places)

In this section you will extend your adding and subtracting skills from 2nd Level where you solved problems involving up to 2 decimal places.

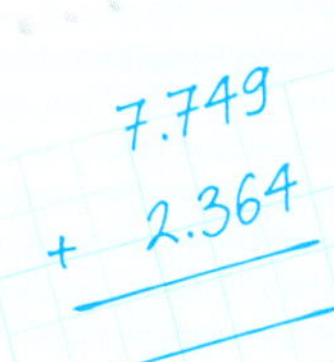

How does that work?

Phillipa has a worktop which is 2·314 metres long.

The space that she wants to fit the worktop into is 1·832 metres long.

How much does Phillipa need to cut from the worktop so that it fits the space?

SOLUTION

- Read the question carefully.
- Strategy: Where have I seen this before?
- Operation: 'Cut from' implies a subtraction (take away sum).
- Answer: This is going to be a measurement.

```
 1 12 11
  2.314
– 1.832
  0.482
```

Phillipa must cut 0·482 metres from the worktop

Be careful if the sum is written out in a row – especially if there are different numbers of decimal places!

Calculate 15·32 + 123·492

SOLUTION

Set out in a column	15·32 +123·492	so that decimal points are under one another That way the digits are also in the correct place
Write in the decimal point	·	
Then add as normal	138·812	

Classroom challenge

Now answer these questions.

Make sure you set out your working clearly.

1 a 7·749 + 2·364

b 5·432 + 6·817

c 0·528 + 4·455

d 8·07 + 3·789

e 6·706 + 12·43

f 0·12 + 4·999

g 5·145 + 3·567 + 3·212

h 0·187 + 6·733 + 1·170

i 2·875 + 0·98 + 3·5

2 a
```
  7·749
– 2·364
```
b
```
  6·817
– 5·432
```
c
```
  4·455
– 0·52
```
d
```
  8·07
– 3·789
```
e
```
 12·46
– 6·706
```
f
```
  4·999
– 0·012
```

When doing adding or subtracting sums, always write out your working in columns.

If there is more than one operation, do them separately.

3 a
```
  7·749
+ 2·364
– 1·567
```
b
```
  4·748
+ 5·303
– 2·117
```
c
```
  8·653
– 4·284
+ 2·314
```
d
```
  3·142
+ 2·782
– 1·005
```
e
```
  6·777
– 2·888
+ 3·444
```
f
```
  5·264
– 0·123
+ 4·168
```

4 Calum and Grigor are buying their mum a Mother's Day present. Calum has saved up £15·86 and Grigor has saved up £23·78. How much have they saved in total?

5 Rory is picking apples on a farm. In the morning he picked 47·315 kg of apples. In the afternoon he picked 36·75 kg. What weight of apples had he picked in total?

6 Patricia bought a box of chocolates. In the box there was 0·876 kg of dark chocolate and 0·914 kg of milk chocolate. What weight of chocolate, in total, was in the box?

7 Johnny went for a training run of 6·135 km. The next day he went for another run of 8·816 km. How far had Johnny run in total?

8 Alison went for a cycle ride and recorded her ride on 'Strava'. The first segment was recorded as 15·250 km. The second segment was recorded as 21·675 km. How far did Alison cycle?

9 Mark measured his height to be 1·83 m. Matthew measured his height to be 1·78 m. How much taller than Matthew is Mark?

10 Claire weighed herself at the start of February when she recorded her weight as 88·09 kg. At the end of February, she weighed herself again and recorded a weight of 84·602 kg. How much weight did Claire lose?

11 Roshan cut a piece of wood 1·435 m long from a plank of length 3 m. How much of the plank was left?

12 A sign on a theme park ride says, 'You must be 140 cm tall'.
Emma is 1·342 m tall.
By how much does Emma miss out on the height restriction?

13 Colin grows marrows to show in competitions.
The longest marrow recorded is 62·45 cm long.
Colin's best marrow is 61·987 cm long.
How much short is he of the record?

14 Usain Bolt ran the Olympic 100 m sprint in London, in 2012, in a time of 9·63 seconds.
In Rio in 2016, he ran the 100 m sprint in 9·81 seconds.
How much faster was he in London?

15 The fastest lap and average speeds of the top five places at the Silverstone Formula 1 event in 2017 are shown in the table.

Position	Driver	Car	Fastest lap min:seconds	Average speed (miles per hour)
1	Lewis Hamilton	Mercedes	1:30·621	234·025
2	Max Verstappen	Red Bull / Tag Heuer	1:30·678	233·878
3	Valtteri Bottas	Mercedes	1:30·905	233·924
4	Kimi Raikkonen	Ferrari	1:31·517	231·733
5	Sebastian Vettel	Ferrari	1:31·872	230·838

a What was the time difference between Raikkonen and Hamilton?
b How many seconds slower was Bottas than Verstappen?
c What was the difference in time between 1st and 5th place?
d How much quicker, in mph, was Hamilton than Bottas?
e How much slower, in mph, was Raikkonen than Verstappen?
f What was the difference in average speed from 1st place to 5th place?
g Lewis Hamilton won the race in a total time of 1 hour 21 minutes and 27·430 seconds.
Valtteri Bottas was +14·063 seconds. What was his total time?
Kimi Raikkonen was +36·570 seconds. What was his total time?

Multiplying and dividing whole numbers and decimals (up to 3 decimal places)

In this section, you will extend your multiplying and dividing skills from 2nd Level where you solved problems involving up to 2 decimal places.

How does that work?

When multiplying a decimal by a whole number, you should set the sum out like this:

When multiplying 4·248 × 7

```
    4·248  ← 3 digits after decimal point
×       7
= 29·736   ← 3 digits after decimal point
```

When multiplying 0·014 × 6

```
    0·014  ← 3 digits after decimal point
×       6
=   0·084  ← 3 digits after decimal points
```

Use 0s as 'place keepers' for the decimal point

Maria is cutting 3 pieces of wood, each 0·755 m long, from a plank that is 3 metres long.

How much wood is left?

SOLUTION

3 × 0·755

```
    0·755  ← 3 digits after decimal point
×       3
=   2·265  ← 3 digits after decimal point
```

Wood left

```
    3·000
–   2·265
=   0·735 m
```

There is 0·735 m of wood left.

EXAMPLE

The weight of a small car is 1·67 tonnes.

What would be the total weight of 300 such cars?

SOLUTION

You need to calculate 1·67 × 300

Here you would do
1·67 × 3 then × 100

That is
5·01 × 100 = 501 tonnes

DON'T FORGET

When multiplying by a multiple of 10, 100, 1000 etc. it is best to split the number. For example

× 40 is the same as × 4 then × 10
× 600 is the same as × 6 then × 100

DON'T FORGET

When multiplying, or dividing, by 10, 100, 1000 etc. remember that the digits move the same number of places as the zeros.

That is, × 10 digits move 1 place to left
× 100 digits move 2 places to left
÷ 1000 digits move 3 places to the right

Classroom challenge

1 a 4.312×4 b 6.123×7 c 8.246×8 d 7.438×9

2 a 16.368×5 b 12.44×6 c 123.7×3 d 8.998×2

3 a 2.15×20 b 3.26×30 c 4.25×60 d 2.91×80

e 21.64×200 f 3.14×500 g 6.18×300 h 62.841×400

DON'T FORGET

Remember to set out your calculation like this:

$0.3 \times 7 = 2.1$ $0.64 \times 9 = 5.76$

For these questions, use the same strategy as you did for the adding and subtracting section.

That is,

- READ the question carefully.
- Select a STRATEGY (or how you are going to tackle the question).
- Decide on the OPERATION(S) to use – is it add, subtract, multiply, divide or a mixture?
- Decide what the ANSWER is going to look like.

DON'T FORGET

You will have learned a number of strategies for solving problems. For example, set out a calculation, make a table, draw a diagram, simplify the problem, make a list, work backwards, look for a pattern.

4 HMS Neptune, a Royal Navy submarine is taking delivery of torpedoes.
Each torpedo weighs 3·25 tonnes.
What is the total weight of 8 torpedoes?

5 Eskbank Cricket Club has a scoring average of 5·67 runs per over.
There are 12 overs left.
At this rate, will Eskbank Cricket Club score the 65 runs needed to win the match?

6 A pleasure boat weighs 5·366 tonnes.
How much will a fleet of 6 pleasure boats weigh?

7 Rosie buys 5 containers of coconut milk.
If each container weighs 1·125 kg, what is the total weight of the containers?

8 The table shows the average weights, in tonnes, of some cars and vans.

Type of vehicle	Weight (in tonnes)
Compact car	1·354
Medium size car	1·590
Large car	1·985
SUV or 4×4 car	1·577

What would be the weight of:

a 7 compact cars
b 6 medium size cars
c 4 SUVs
d 5 large cars
e 400 compact cars
f 3000 large cars
g 7000 medium size cars?

9 A bag of flour has a label on it stating: 'Weight 1 lb/0·454 kg'.
a What would 12 bags of flour weigh in kilograms?
b A warehouse buys 5000 bags of flour. What would be the total weight of flour in kilograms?

10 1 mile per hour is equivalent to 0·871 knots (the unit in which the speed of a boat is measured).
The speed limit for a narrow boat on a canal is 4 miles per hour.
How many knots is this?

11 A plane is flying at a speed of 'mach 1'. This is equivalent to 761·213 miles per hour.
How fast, in miles per hour, would a plane be travelling if it was at 'mach 3'?

How does that work?

When dividing a decimal by a whole number, you should set out the sum like this:

Look at the calculation $7\overline{)1\,7\cdot2\,2}$

What you are writing	What you are thinking
$7\overline{)1\,7\cdot2\,2}$	7 into 1 does not go => carry to 7
$\begin{array}{r} \mathbf{2}\cdot \\ 7\overline{)1\,{}^{1}7\cdot2\,2} \end{array}$	7 into 17 goes 2 times, remainder 3
$\begin{array}{r} 2\cdot\mathbf{4} \\ 7\overline{)1\,{}^{1}7\cdot{}^{3}2\,2} \end{array}$	7 into 32 goes 4 times, remainder 4
$\begin{array}{r} 2\cdot4\,\mathbf{6} \\ 7\overline{)1\,{}^{1}7\cdot{}^{3}2\,{}^{4}2} \end{array}$	7 into 42 goes 6 times

The answer is 2·46

When dividing by a multiple of 10, 100, 1000 etc. you use a similar method as for multiplying.

That is, split the number and then use the technique as above.

EXAMPLE

If you want to divide 17·22 by 700 the steps to follow would be:

17·22 ÷ 700	is the same as	17·22 ÷ 7 then	÷ 100
17·22 ÷ 7	is the example above, giving	2·46	÷ 100
The answer is		0·0246	

Classroom challenge

1 a 2·146 ÷ 2 b 3·183 ÷ 3 c 5·488 ÷ 4 d 5·616 ÷ 8
e 15·525 ÷ 5 f 8·109 ÷ 9 g 1·818 ÷ 9 h 3·545 ÷ 5

2 a 36·22 ÷ 20 b 4·97 ÷ 70 c 7·55 ÷ 50 d 3·72 ÷ 60
e 1·28 ÷ 40 f 3511 ÷ 500 g 2807 ÷ 700 h 4848 ÷ 200

3 6 bags of sugar weigh 13·2 lb.
What is the weight of 1 bag of sugar?

4 A fleet of 8 identical hire cars weighs 10·832 tonnes.
What is the weight of one car?

5 Karen travels 63·875 kilometres on 7 litres of fuel.
How far would she travel on 1 litre of fuel?

6 The total length of 4 identical shelves is 4·5 metres.
What is the length of each shelf?

7 9 lbs is equivalent to 4·086 kilograms.
What is 1 lb equivalent to?

8 A rope is 24·48 metres long.
It is cut into 5 equal pieces.
How long is each piece?

9 A baker mixes 1·665 kg of dough.
From this he makes 9 round loaves.
What weight of dough is in each loaf?

10 A teacher buys a class set of 30 textbooks at a total cost of £659·70.
What is the cost of one textbook?

11 The total weight of 3000 cars is 4956 tonnes.
What is the weight of one car?

12 The total weight of 6000 packets of spice powder is 1·524 kg.
What is the weight, in grams, of one packet of spice powder?

Working with integers

“I can use my understanding of numbers less than zero to solve simple problems in context.” MNU 3-04a

What’s coming up?

This Outcome and Experience will give you the opportunity to:

- solve addition and subtraction problems working with integers
- solve multiplication and division problems working with integers.

What you already know

You have already learned how to:

✔ identify familiar contexts in which negative numbers are used
✔ order numbers less than zero and locate them on a number line.

There are many areas in real life where you will have to work with integers, including negative numbers. For example, in temperature, heights above or below sea level, and when working with bank accounts.

Adding and subtracting integers

How does that work?

Consider this number pattern:

$4 + 3 = 7$	Think about how you move along the number line	
$4 + 2 = 6$	As you reduce the adding number by 1 so the answer reduces by 1	
$4 + 1 = 5$		
$4 + 0 = 4$		
$4 + (-1) = 3$	same as $4 - 1$	so adding a negative is same as subtracting
$4 + (-2) = 2$	same as $4 - 2$	
$4 + (-3) = 1$	same as $4 - 3$	

What is $6 + (-2)$
Same as $6 - 2$
$= 4$

What is $(-4) + (-3)$
Same as $(-4) - 3$
$= (-7)$

Brackets are often put round negative numbers to separate them from the subtraction (minus) sign.

Consider this number pattern

4 − 3 = 1			
4 − 2 = 2	as you reduce the subtracting number by 1 so the answer increases by 1		
4 − 1 = 3			
4 − 0 = 4			
4 − (−1) = 5	same as	4 + 1	so subtracting a negative is the same as adding
4 − (−2) = 6	same as	4 + 2	
4 − (−3) = 7	same as	4 + 3	

What is 6 − (−2)
Same as 6 + 2
= 8
What is (−4) − (−3)
Same as (−4)+3
= (−1)

Classroom challenge

1 A quick reminder about negative numbers around you.
  a Put these in order, starting from smallest. 5°C, −3°C, −1°C, 3°C
  b Put these in order, starting from the smallest. −4°C, 0°C, 2°C, −1°C, 4°C
  c How many degrees Celsius are there between. −4°C and 5°C
  −8°C and −3°C
  15°C and −5°C?
  d Marc has £20 in his bank account. He takes out £15.
  What is his new balance?
  He withdraws a further £10. What will his balance show now?
  e Badwater Basin, in California is 85 metres below sea level (−85 m).
  The height of Ben Nevis, in Scotland, is 1345 metres (1345 m) above sea level.
  What is the difference in metres between the two?

2 Complete these number patterns.

Set 1	Set 2	Set 3
6 + 3 =	8 + 3 =	7 + 3 =
6 + 2 =	8 + 2 =	7 + 2 =
6 + 1 =	8 + 1 =	7 + 1 =
6 + 0 =	8 + 0 =	7 + 0 =
6 + (−1) =	8 + (−1) =	7 + (−1) =
6 + (−2) =	8 + (−2) =	7 + (−2) =
6 + (−3) =	8 + (−3) =	7 + (−3) =

3 Calculate:

a $7 + (-3)$ b $8 + (-2)$ c $9 + (-5)$ d $5 + (-6)$
e $(-4) + 1$ f $(-5) + (-2)$ g $(-3) + (-5)$ h $(-6) + (-6)$

4 Work out the answer to:

a $15 + (-8)$ b $23 + (-11)$ c $40 + (-20)$ d $100 + (-80)$
e $(-25) + 25$ f $(-400) + 250$ g $(-65) + (-5)$ h $(-250) + (-250)$

5 Now try these:

a $4 + 3 + (-2)$ b $7 + 10 + (-6)$ c $(-4) + 6 + (-2)$ d $(-5) + (-3) + 10$
e $22 + 11 + (-33)$ f $40 + 30 + (-25)$ g $(-20) + (-20) + 2$ h $(-5) + (-6) + (-7)$

6 Complete these number patterns.

Set 1	Set 2	Set 3
6 − 3 =	8 − 3 =	7 − 3 =
6 − 2 =	8 − 2 =	7 − 2 =
6 − 1 =	8 − 1 =	7 − 1 =
6 − 0 =	8 − 0 =	7 − 0 =

Set 1	Set 2	Set 3
6 − (−1) =	8 − (−1) =	7 − (−1) =
6 − (−2) =	8 − (−2) =	7 − (−2) =
6 − (−3) =	8 − (−3) =	7 − (−3) =

7 Calculate:

a $7 - (-3)$ b $8 - (-2)$ c $9 - (-5)$ d $5 - (-6)$
e $(-4) - 1$ f $(-5) - (-2)$ g $(-3) - (-5)$ h $(-6) - (-6)$

8 Work out the answer to:

a $15 - (-8)$ b $23 - (-11)$ c $40 - (-20)$ d $100 - (-80)$
e $(-25) - 25$ f $(-400) - 250$ g $(-65) - (-5)$ h $(-250) - (-250)$

9 Now try these:

a $4 + 3 - (-2)$ b $7 + 10 - (-6)$ c $(-4) + 6 - (-2)$ d $(-5) - (-3) + 10$
e $22 + 11 - (-33)$ f $40 + 30 - (-25)$ g $(-20) - (-20) + 2$ h $(-5) - (-6) - (-7)$

10 While checking last month's bank statement, George noticed that he had spent £100 on shopping and had deposited £100 in his account.
Which integer describes the change in George's account?
−£100, £0, £100, £10

11 In a board game, Ronnie scored 50 points in the first half.
In the second half he scored −20 points.
What was Ronnie's total score?
−70, −30, 30, 70

12 Kylie was climbing up a hill – she climbed 500 ft.
The starting point of the climb was 100 ft below sea level.
Which integer represents her height now?
400 ft, −400 ft, 600 ft, −600 ft

13 The boiling point of nitric acid is 83°C.
The freezing point of nitric acid is – 42°C.
What is the difference, in degrees, between the boiling and freezing points of nitric acid?

Multiplying and dividing integers

Multiplying is a quick way of repeatedly adding the same number.

How does that work?

4 × 5	is 4 'lots' of 5	that is	5	+ 5	+ 5	+ 5	= 20
4 × (–5)	is 4 'lots' of (–5)	that is	(–5)	+(–5)	+(–5)	+(–5)	= –20
–4 × 5	is –4 'lots' of 5	that is	–(5)	+–(5)	+–(5)	+–(5)	
		which gives	– 5	– 5	– 5	– 5	= –20
–4 × (–5)	is –4 'lots' of (–5)	that is	–(–5)	–(–5)	–(–5)	–(–5)	
		which gives	5	+ 5	+ 5	+ 5	= 20

This gives us a 'rule of signs' for multiplying integers

+ + }		
– – }	Same sign gives a positive	+ve
+ – }		
– + }	Opposite signs give a negative	–ve

How does that work?

When doing a multiplication sum, do it in 2 steps:

- ✔ Determine sign
- ✔ Determine number.

So, for example, (–6) × 4

- ✔ Opposite signs so – ve
- ✔ 6 × 4   24   so answer is –24

Another one, (–6) × (–4)

- ✔ Same signs so + ve
- ✔ 6 × 4   24   so answer is +24   (normally we would simply write 24)

As dividing is the reverse process to multiplying we can use the same rule.
So, for example, (–12) ÷ 4

- ✔ Opposite signs so – ve
- ✔ 12 ÷ 4   3   so answer is –3

Another one, (–12) ÷ (–4)

- ✔ Same signs so + ve
- ✔ 12 ÷ 4   3   so answer is +3   (normally we would simply write 3)

Classroom challenge

1 Copy and complete these multiplication sums.

Set 1	Set 2	Set 3
6 × (–3)	(–4) × 7	(–4) × (–5)
8 × (–5)	(–5) × 5	(–3) × (–6)
10 × (–4)	(–8) × 9	(–8) × (–6)
4 × (–7)	(–12) × 6	(–12) × (–11)
9 × (–5)	(–11) × 4	(–9) × (–3)

2 Copy and complete these multiplication squares.

a

×	2	–4	–7	5
2				
6				
–4				
–8				

b

×	3	7	–8	–6
12				
–5				
–6				
10				

3 Copy these calculations, and fill in the missing numbers.

a ..... × (–5) = –20 b ..... × 6 = –42 c (–5) × ..... = 35

d 12 × ..... = –144 e (–11) × ..... = 110 f (–8) × ..... = 64

4 Calculate:

a 4 × (–3) × (–2) b (–6) × 4 × (–2) c 6 × 2 × (–5)

d (–3) × (–2) × (–4) e (–8) × (–3) × 2 f (–4) × (–4) × (–4)

5 Yesterday's lowest temperature was –2°C.
Today it is 4 times lower.
What is today's lowest temperature?

6 Copy and complete this table in two different ways.

×				
	1			
		4		
			9	
				16

DON'T FORGET

6 × 6 = 36
(–6) × (–6) = 36

7 Copy and complete these division sums.

Set 1	Set 2	Set 3
6 ÷ (–3)	(–4) ÷ 2	(–4) ÷ (–4)
8 ÷ (–2)	(–5) ÷ 1	(–6) ÷ (–3)
10 ÷ (–5)	(–8) ÷ 4	(–8) ÷ (–2)
4 ÷ (–4)	(–12) ÷ 3	(–11) ÷ (–11)
9 ÷ (–3)	(–11) ÷ 11	(–9) ÷ (–3)

8 Calculate:

a $(-10) \div (-2)$ b $(-15) \div 5$ c $18 \div (-3)$ d $14 \div (-7)$
e $(-45) \div 9$ f $(-20) \div (-10)$ g $(-100) \div (-4)$ h $(-25) \div (-5)$
i $80 \div (-4)$ j $(-144) \div (-12)$ k $121 \div (-11)$ l $(-42) \div (-7)$

9 Copy these calculations, and fill in the missing numbers.

a $(-40) \div ..... = -8$ b $(-35) \div ..... = -5$ c $(-80) \div ..... = 4$
d $..... \div (-6) = -42$ e $..... \div (-15) = 2$ f $100 \div ..... = -25$

10 The temperature in a room dropped by 2°F every hour for 5 hours.
What was the change in temperature over the 5 hours?

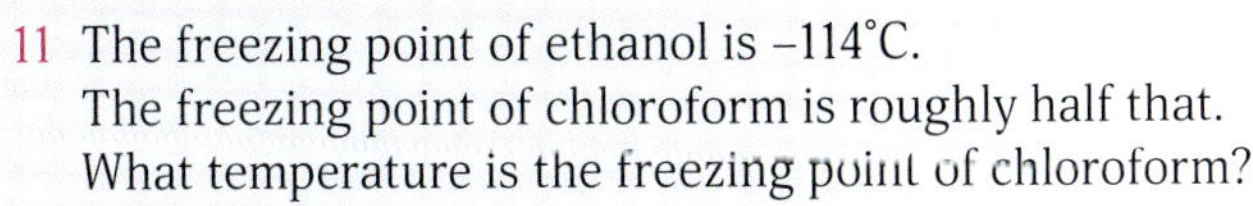

11 The freezing point of ethanol is –114°C.
The freezing point of chloroform is roughly half that.
What temperature is the freezing point of chloroform?

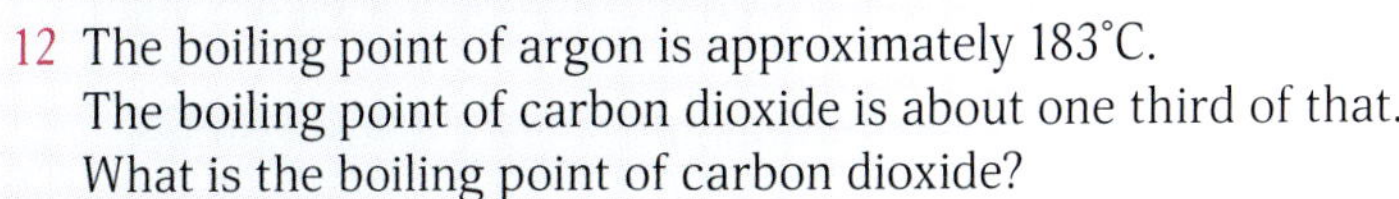

12 The boiling point of argon is approximately 183°C.
The boiling point of carbon dioxide is about one third of that.
What is the boiling point of carbon dioxide?

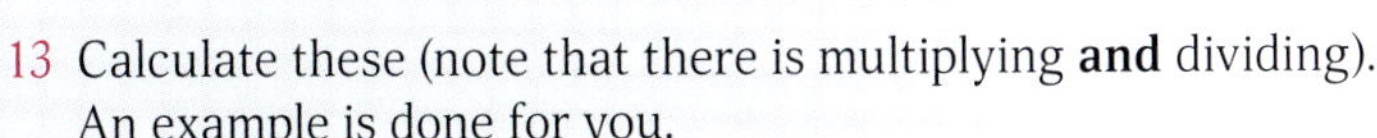

13 Calculate these (note that there is multiplying **and** dividing).
An example is done for you.

$$\frac{(-3) \times (-4)}{(-2)}$$

Simplify the numerator	$(-3) \times (-4)$	=	12
Denominator already simplified		=	(-2)
	$12 \div (-2)$	=	-6

a $\dfrac{(-4) \times (-6)}{12}$ b $\dfrac{5 \times (-6)}{(-15)}$ c $\dfrac{(-8) \div 2}{4}$ d $\dfrac{(-14) \div (-2)}{(-1)}$

e $\dfrac{(-7) \times (-5) \times (-2)}{5}$ f $\dfrac{8 \times 6 \times (-9)}{(-2) \times (-3)}$ g $\dfrac{(-4) \times (-7) \times 3}{(-12) \div (-2)}$

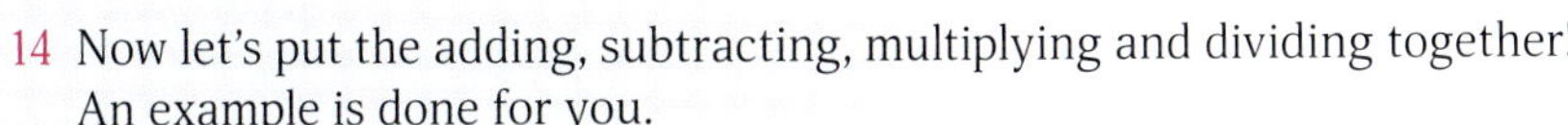

14 Now let's put the adding, subtracting, multiplying and dividing together!
An example is done for you.

$$\begin{aligned} & (-2) \times (-5) + 4 \times (-3) \\ &= 10 + (-12) \\ &= -2 \end{aligned}$$

a $(-6 + 10) \div (-2)$ b $(12 - 24) \div (-6)$ c $(6 + (-8)) \times (2 + (-1))$

d $(-4 \times -3) + (16 \div (-8))$ e $(8 \div (-2)) - (3 \times (-4))$ f $\dfrac{((-3) \times (-4)) + (2 \times (-1))}{(-5)}$

MULTIPLES, FACTORS AND PRIMES

Common multiples and factors

I have investigated strategies for identifying common multiples and common factors, explaining my ideas to others, and can apply my understanding to solve related problems. MTH 3-05a

What's coming up?

This Outcome and Experience will give you the opportunity to:

- identify common multiples, including the lowest common multiple for whole numbers and explain the method used
- identify common factors, including the highest common factor for whole numbers and explain the method used
- solve problems using multiples and factors.

What you already know

You have already learned how to:

- ✔ identify multiples and factors of whole numbers
- ✔ apply knowledge and understanding of these when solving relevant problems in number, money and measurement.

Why do we need to learn how to find multiples, lowest common multiples (LCMs), or highest common factors (HCFs)?

Easy! So we can do what mathematics is all about: solving problems!

Common multiples

How does that work?

Multiples are simply your 'times tables'.

For example:

Write down the first twelve multiples of 3
They are 3, **6**, 9, **12**, 15, **18**, 21, **24**, 27, 30, 33, 36

Write down the first twelve multiples of 2
They are 2, 4, **6**, 8, 10, **12**, 14, 16, **18**, 20, 22, 24
Do you notice any multiples in common? 6, 12, 18, 24

What is the lowest number in this list? 6

6 is the **lowest common multiple (LCM)** of 2 and 3

The lowest common multiple is useful when working out real-life problems.

How does that work?

EXAMPLE

Katrina is having a barbeque.
She goes to the shop to buy burgers and buns.
The burgers come in packs of 6, but the buns come in packs of 8.
What is the least number of burgers and buns Katrina could buy to make sure there are no 'extras' or waste?

SOLUTION

So, Katrina could have:

burgers	6	12	18	**24**	30	36	...
buns	8	16	**24**	32	40	48	...

Katrina sees that 24 is the lowest common multiple (LCM).

So she should buy 4 packs of burgers = 4×6 = 24 burgers
3 packs of buns = 3×8 = 24 buns

That is, each bun has a burger and there are no leftover buns or burgers!

How does that work?

The lowest common multiple is also really useful when adding, or subtracting, 'unlike' fractions.

We can 'multiply up' to give a common denominator.

For example,

$\frac{1}{2} + \frac{1}{3}$ The LCM of 2 and 3 is 6 Multiples of 2 2, 4, **6**, 8, 10 ...

Multiples of 3 3, **6**, 9, 12 ...

$= \frac{3}{6} + \frac{2}{6}$

$= \frac{5}{6}$

DON'T FORGET

We will revisit this later in the book, in the 'fractions' section.

Classroom challenge

1 Write down:

a the first twelve multiples of 4

b the first nine multiples of 3

c the first eight multiples of 5

d the first ten multiples of 10.

2 Write down:

a all the multiples of 4 between 31 and 61

b all the multiples of 9 between 26 and 96

c all the multiples of 7 between 40 and 100

d all the multiples of 6 between 50 and 110.

3 a Write down the first fifteen multiples of 2.

b What is the name given to these numbers?

4 21, 24, 27, 30, 33, 36 could be described as 'the multiples of 3 from 21 to 36'.
Describe the following lists of numbers in the same way.

a 15, 20, 25, 30, 35, 40

b 60, 66, 72, 78, 84

c 36, 45, 54, 63, 72, 81, 90

d 70, 80, 90, 100, 110, 120

e 60, 80, 100, 120, 140

f 42, 45, 48, 51, 54

5 Find the LCM of the following pairs of numbers.

a 3 and 4

b 2 and 5

c 6 and 9

d 4 and 10

e 8 and 7

f 6 and 5

g 7 and 3

h 5 and 12

i 7 and 11

DON'T FORGET

You may wish to write out the first few multiples of the larger number – and look for the first number in the list into which the smaller number divides exactly.

6 a Write down the first ten multiples of 2.

b Write down the first eight multiples of 3.

c Write down the first six multiples of 9.

d Write down the lowest common multiple (LCM) of 2, 3 and 9.

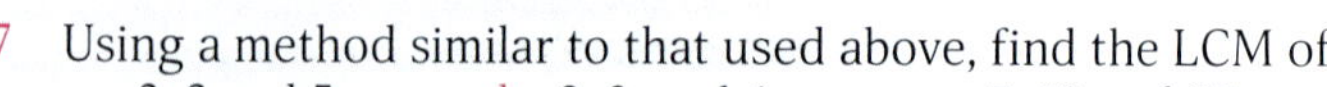

7 Using a method similar to that used above, find the LCM of:

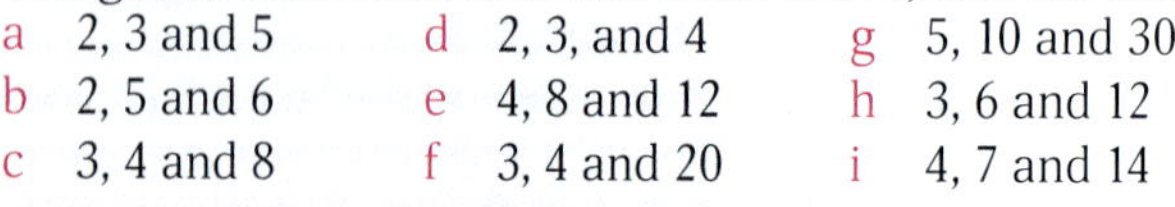

a 2, 3 and 5

b 2, 5 and 6

c 3, 4 and 8

d 2, 3, and 4

e 4, 8 and 12

f 3, 4 and 20

g 5, 10 and 30

h 3, 6 and 12

i 4, 7 and 14

8 Graeme is a lighthouse keeper.
He sets the light to flash every 3 minutes.
He sets the horn to sound every 7 minutes.
At 9pm, both the light flashes and the horn sounds.
When will they next flash and sound together?

DON'T FORGET

You should consider writing out the multiples of 3 and then the multiples of 7. Is there a common multiple? What is the LCM?

9 In the United States I can vote for a president every 4 years in a ballot.
I can vote for the Senate every 6 years in a ballot.
In 2018, I voted for both the President and the Senate in the same ballot.
In what year will I, again, be able to vote for both of these positions in the same ballot?

10 The number 3 tram takes 24 minutes to complete its route and return to the terminus.
The number 7 tram takes 18 minutes to complete its route and return to the terminus.
Both trams leave the terminus at 8am.
When will they next be at the terminus together?

11 Marcus is setting up his Christmas lights.
The red lights flash every 6 seconds
The blue lights flash every 9 seconds
The yellow lights flash every 10 seconds.
Marcus notices that all three lights have flashed together.
After how many seconds will the lights flash together again?

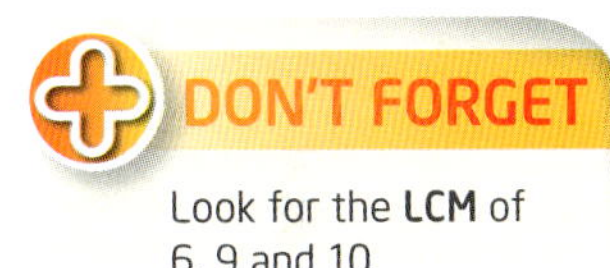

Look for the **LCM** of 6, 9 and 10.

12 Jenny pays her telephone bill every 8 weeks.
She pays her electricity bill every 9 weeks.
She pays her gas bill every 6 weeks.
Jenny notices that she paid them all on 1st June.
After how many weeks will she next pay them all at the same time?

13 Callum, Grigor and Rory live in Australia and use 'Skype' to talk to their mother in Scotland.
Callum calls every 8 days.
Grigor calls every 5 days.
Rory calls every 6 days.
Today, all three called their mother.
How many days will it be until all three call on the same day again?

STRETCH YOURSELF

14 The Earth takes 1 year to orbit the Sun.
Venus takes $\frac{2}{3}$ of a year to orbit the Sun.
If the two planets are in alignment, as shown in the diagram, how long will it be before they return to this position?

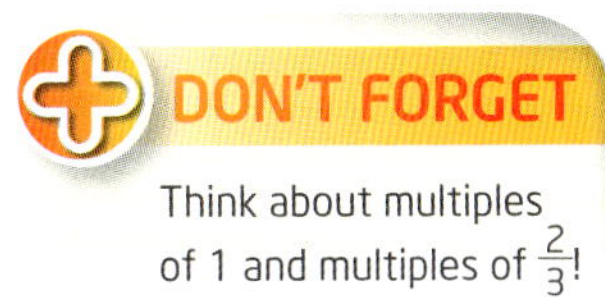

Think about multiples of 1 and multiples of $\frac{2}{3}$!

STRETCH YOURSELF

15 Now, let's add in Mercury!
Mercury takes $\frac{1}{4}$ of a year to orbit the Sun.
Remember: Earth takes 1 year.
Venus takes $\frac{2}{3}$ of a year.
All three are in alignment with the Sun.
How many years will it take for all three to return to this alignment?

Common factors

Remember that factors are simply numbers that divide exactly into another number.

For example, the factors of 6 are 1, 2, 3, 6.

How does that work?

EXAMPLE

Write down all the factors of 12.

DON'T FORGET

The factors of a number will always include 1 and the number itself.

SOLUTION

It is best to work in 'number pairs' to find factors.

12 = 1 × 12
2 × 6
3 × 4

We can stop here as the next 'pair' would be 4 × 3 which is same as 3 × 4.

The factors of 12 are 1, 2, 3, 4, 6, 12.

Like multiples, factors are often used in fraction calculations to simplify the fraction.

Common factors also appear in algebra.

However, common factors, and especially the **highest common factor (HCF)**, are often used to solve problems.

How does that work?

EXAMPLE

Clarke is catering for a party.
He has 72 sticks of celery and 48 carrot sticks.
He wants both foods to be on each plate.
He wants to share the food out evenly and have no waste.
What is the largest number of plates he can make up?
How much of each food would be on each plate?

SOLUTION

Write down all the factors of 72. Write down all the factors of 48.
When writing down factors it is best to write them in pairs – so that you don't miss any out!

72		
1	72	1 × 72
2	36	2 × 36
3	24	3 × 24
4	18	4 × 18
6	12	6 × 12
8	9	8 × 9

48		
1	48	1 × 48
2	24	2 × 24
3	16	3 × 16
4	12	4 × 12
6	8	6 × 8

It is now easy to list the factors, in order, for each number.

Factors of 72 are **1**, **2**, **3**, **4**, **6**, **8**, 9, **12**, 18, **24**, 36, 72
Factors of 48 are **1**, **2**, **3**, **4**, **6**, **8**, **12**, 16, **24**, 48
Factors common to both lists are **1**, **2**, **3**, **4**, **6**, **8**, **12**, **24**
The highest number in this list is 24
The **highest common factor (HCF)** of 72 and 48 is **24**
This means that the greatest number of plates Clarke could have is **24**

For food on each plate:
celery sticks 72 ÷ **24** = 3 so 3 sticks of celery on each plate.
carrot sticks 48 ÷ **24** = 2 so 2 sticks of carrot on each plate.

Classroom challenge

1 Write down all four factors of 8. List them in order from smallest to largest.
2 Write down all six factors of 18. Put them in order from smallest to largest.
3 The number 10 has four factors. List them in order, smallest to largest.
4 Write down all three factors of 25. List them in order from smallest to largest.
5 Write down all the factors of 36. List them in order from smallest to largest.
6 Using 'the pairs method' write down all the factors of:

a 7	d 22	g 32	j 45
b 14	e 40	h 100	k 23
c 24	f 29	i 54	l 60

7 What do you notice about the number of factors for each part of question 6?
8 Write down all the factors of:

a 1	d 16	g 49	j 100
b 4	e 25	h 64	k 121
c 9	f 36	i 81	l 144

9 a. What do you notice about the number of factors for each part of question7?
b. What do we call this type of number? (1, 4, 9, 16, 25, 36 … …)

10 Write down all the factors of:

a 2	d 7	g 17	j 29
b 3	e 11	h 19	k 31
c 5	f 13	i 23	l 37

11 a. What do you notice about the number of factors for each part of question 9?
b. Do you know what these numbers are called?
We will look at them more closely in the next section.

12 a. List all the factors of 12, in order from smallest to largest.
b. List all the factors of 8, in order from smallest to largest.
c. Write all the common factors; that is the numbers which appear in both lists.
d. What is the highest common factor (HCF) of 12 and 18?

13 a. List, in order, all the factors of 15.
b. List, in order, all the factors of 20.
c. What is the HCF of 15 and 20?

14 Find the highest common factor (HCF) for each of the following pairs of numbers.

a 6 and 8	c 12 and 24	e 15 and 60	g 19 and 57
b 30 and 50	d 24 and 36	f 8 and 36	h 18 and 24

15 Alicia is working for her school charities group.
She has 36 pencils and 24 pens which she is putting into pencil cases to send to children in need.
She wants to share the pencils and pens equally, and does not want any left over.
a What is the maximum number of pencil cases Alicia could fill?
b How many pencils and pens would go into each pencil case?

Look back at the example.

16 Elton works in a florist shop.
He has 48 red tulips and 32 white tulips.
He wants to share the red and white tulips equally in vases.
He does not want to have any tulips left over.
a What is the maximum number of vases Elton could make?
b How many of each colour of tulip would go into each vase?

17 George is trying to encourage his friends to use public transport.
He has 18 bus tickets and 12 tram tickets.
He wants to make up envelopes to give to his friends.
He wants the envelopes to contain both bus and tram tickets, and for these to be shared equally.
He does not want any tickets left over.
a What is the maximum number of envelopes George could fill?
b How many bus tickets and how many tram tickets would be in each envelope?

18 Lorna is making bracelets using beads.
She has 45 green beads and 36 blue beads.
She wants to share the green and blue beads equally on the bracelets.
She does not want to have any beads left over.
a What is the maximum number of bracelets Lorna could make?
b How many of each colour of bead would be on each bracelet?

STRETCH YOURSELF

19 Lorna has bought in new stock of beads.
She now has 120 green beads
90 blue beads
45 white beads.
She wants to share the 3 colours on all the bracelets.
She does not want any beads left over.
a What is the maximum number of bracelets Lorna could make?
b How many of each colour of bead would be on each bracelet?

Download the Homework Helpers from the Bright Red website for an additional exercise on this topic!

Working with prime numbers

I can apply my understanding of factors to investigate and identify when a number is prime. MTH 3-05b

What's coming up?

This Outcome and Experience will give you the opportunity to:
- identify prime numbers to 100 and explain the method used
- write a given number as a product of its prime factors.

What you already know

You have already learned how to:
- ✔ identify multiples and factors of whole numbers
- ✔ apply knowledge and understanding of these when solving relevant problems in number, money and measurement.

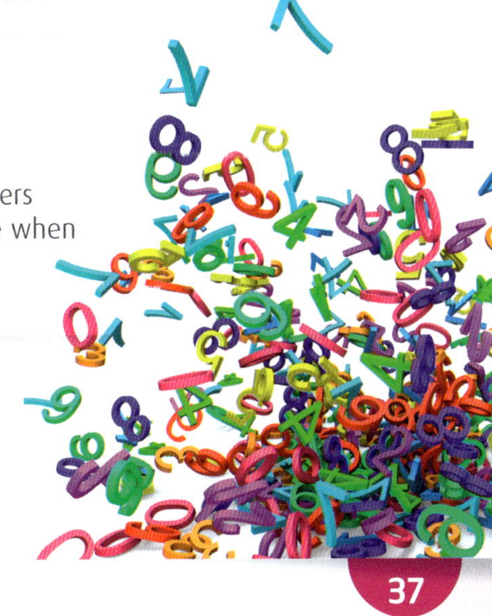

Classroom challenge

What is a prime number?
'A prime number is a number that has exactly one pair of factors.'

1 Finding the prime numbers up to 100.
Get a 1–100 square grid.

This method of finding prime numbers was developed by a Greek mathematician called Eratosthenes in around 200BC. It is called the 'Sieve of Eratosthenes' as it 'sifts out' non-prime numbers.

1	2	3	4	5	6	7	8	9	10
11	12	13	14	15	16	17	18	19	20
21	22	23	24	25	26	27	28	29	30
31	32	33	34	35	36	37	38	39	40
41	42	43	44	45	46	47	48	49	50
51	52	53	54	55	56	57	58	59	60
61	62	63	64	65	66	67	68	69	70
71	72	73	74	75	76	77	78	79	80
81	82	83	84	85	86	87	88	89	90
91	92	93	94	95	96	97	98	99	100

Use black to cross out the 1.
Put a green box around the 2, and then shade in green all the numbers that can be divided by 2.
The next available number is 3.
Put a red box around the 3, and then shade in red all the numbers that can be divided by 3 (the first one will be $3 \times 3 = 9$, as the 6 is already shaded in).
The next available number is 5.
Put a blue box around the 5, and then shade in blue all the numbers that can be divided by 5 (the first will be $5 \times 5 = 25$, as earlier ones are already shaded in).
The next available number is 7.
Put an orange box around the 7, and then shade in orange all the numbers that can be divided by 7 (the first will be $7 \times 7 = 49$).
That is the top row completed. The numbers not shaded are prime numbers.
Why do we not need to check 11?
Look at the pattern above. The first numbers to shade in were 3×3, 5×5, 7×7.
So, if we did 11×11 giving 121, this is 'outside' the 100 square that we have.
Write the list of all the prime numbers up to 100.
Now that we know some of the prime numbers, we can break other numbers – composite numbers – into prime factors as in the example above.

How does that work?

In the previous section, we worked out factors by using 'factor pairs'.

Another method to find factors, especially prime factors, is to use a factor tree.
To find the prime factors of 12 we set it out like this:

Start with the number.
Choose any two factors.
Are they all prime?
If not, break down non-primes.
So 12 = 2 × 2 × 3

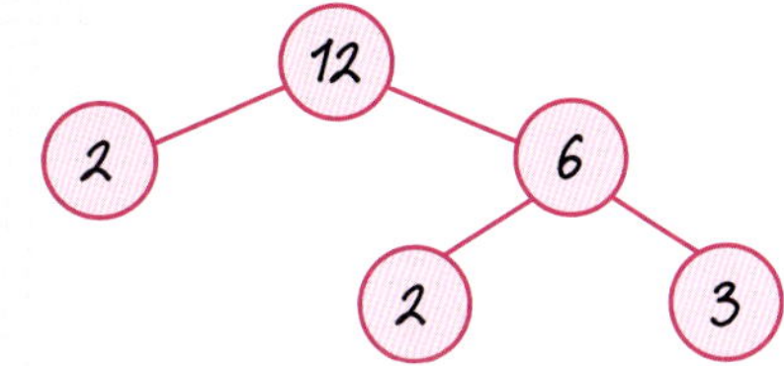

EXAMPLE

Using a factor tree, write down the prime factors of 8.

SOLUTION

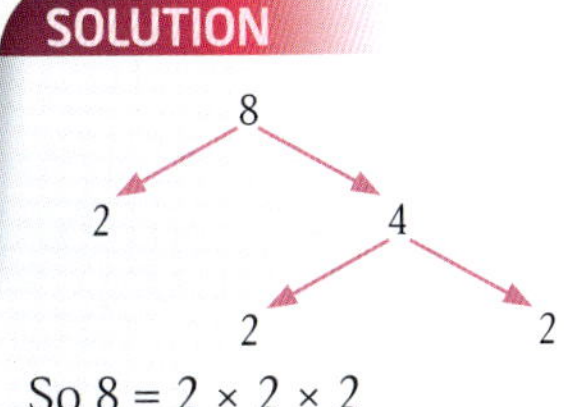

So 8 = 2 × 2 × 2

This is called **prime decomposition**.

2 Using a factor tree write down the prime factors of:

a	10	d	21	g	27	j	72
b	15	e	16	h	30	k	120
c	18	f	45	i	86	l	98

3 Prime decomposition of 48, 4 ways!
There are a number of ways you could draw a factor tree for 48.
Do you always get the same answer?
Copy these out and complete them to check.

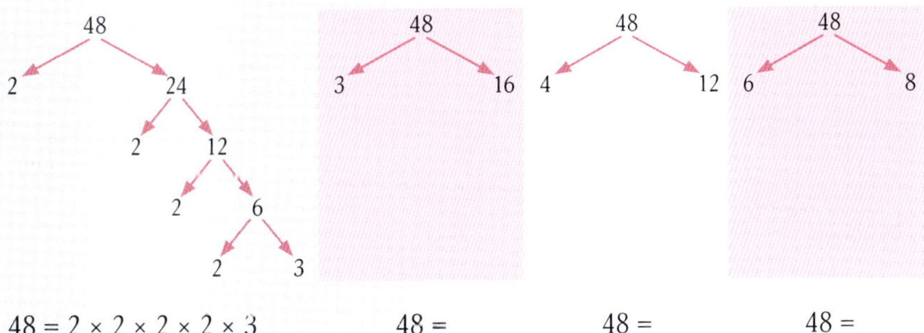

48 = 2 × 2 × 2 × 2 × 3 48 = 48 = 48 =

4 Bill is thinking of a prime number between 40 and 50.
Ben is thinking of a prime number between 10 and 20.
The difference between the two numbers is 26.
What are Bill and Ben's two numbers?

How does that work?

Using prime factors to work out LCM or HCF

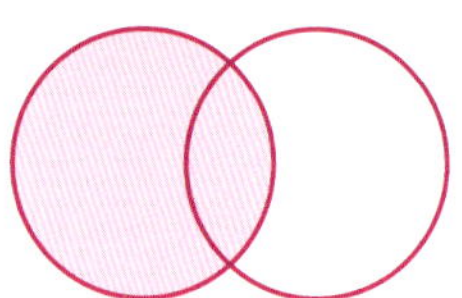

What is the LCM and HCF of 30 and 42?
A useful tool is to use a Venn diagram.
Named after John Venn, the English mathematician.

Firstly, find the prime factors of each number.
30 = 2 × 3 × 5 42 = 2 × 3 × 7
Draw 2 circles (as in a Venn diagram).

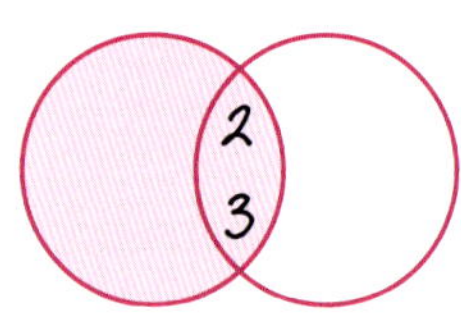

Note that 2 and 3 are in both lists.
They go in the overlapping part.
The remaining numbers go into their part of the diagram.
We end up with a diagram like this:

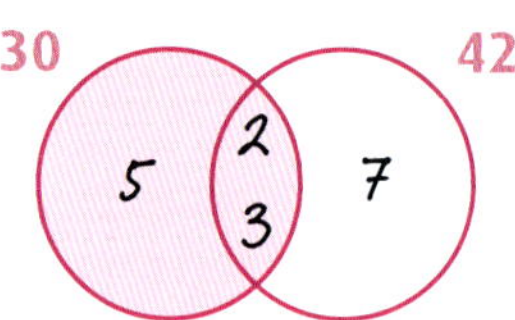

The HCF is 2 × 3 = 6
(numbers in overlap).
The LCM is 5 × 2 × 3 × 7 = 210
(multiply all the numbers together).

5 Using the same method, find the HCF and LCM of 28 and 60.

STRETCH YOURSELF

6 **Reversing primes**

How many 'reversible primes' are there between 10 and 99?
A reversible prime is a prime number which, when the digits are reversed, is still a prime number.
For example, 17 is a reversible prime, as when reversed gives 71 which is also a prime number.
Your challenge is to list them all.

Hints on how to work this out can be found in our Homework Helpers

A game to finish the chapter!

A game for two players 'FACTORED OUT'.

Get a 1–100 grid

Player 1 chooses a number
must be less than 50, and an even number cross it out.
Player 2 chooses a number
must be a multiple or factor of player 1's number cross it out.
Player 2 chooses a number
must be a multiple or factor of player 2's number cross it out.

Continue until a player cannot score out a number.
That player loses; the other wins!

POWERS AND ROOTS

Having explored the notation and vocabulary associated with whole number powers and the advantages of writing numbers in this form, I can evaluate powers of whole numbers mentally or using technology. MTH 3-06a

What's coming up?

This Outcome and Experience will give you the opportunity to:

- explain the notation and use associated vocabulary appropriately, for example index, exponent and power
- evaluate whole number powers, for example $2^4 = 16$
- express whole numbers as powers, for example $27 = 3^3$.

What you already know

Useful learning would include:

- ✓ multiplication tables
- ✓ factors.

Powers

Look at this multiplication square.

Look closely at the numbers in the diagonal, shaded red.

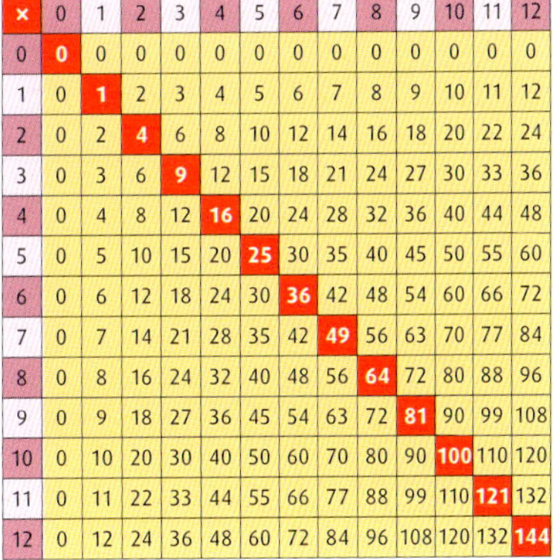

×	0	1	2	3	4	5	6	7	8	9	10	11	12
0	0	0	0	0	0	0	0	0	0	0	0	0	0
1	0	1	2	3	4	5	6	7	8	9	10	11	12
2	0	2	4	6	8	10	12	14	16	18	20	22	24
3	0	3	6	9	12	15	18	21	24	27	30	33	36
4	0	4	8	12	16	20	24	28	32	36	40	44	48
5	0	5	10	15	20	25	30	35	40	45	50	55	60
6	0	6	12	18	24	30	36	42	48	54	60	66	72
7	0	7	14	21	28	35	42	49	56	63	70	77	84
8	0	8	16	24	32	40	48	56	64	72	80	88	96
9	0	9	18	27	36	45	54	63	72	81	90	99	108
10	0	10	20	30	40	50	60	70	80	90	100	110	120
11	0	11	22	33	44	55	66	77	88	99	110	121	132
12	0	12	24	36	48	60	72	84	96	108	120	132	144

How does that work?

The pattern is formed by doing:

$0 \times 0 = 0$
$1 \times 1 = 1$
$2 \times 2 = 2$
$3 \times 3 = 9$
$4 \times 4 = 16$
$5 \times 5 = 25$
$6 \times 6 = 36$
$7 \times 7 = 49$
$8 \times 8 = 64$
$9 \times 9 = 81$
$10 \times 10 = 100$
$11 \times 11 = 121$
$12 \times 12 = 144$

These numbers, 1, 4, 9, 16, 25, and so on, are called **square numbers**. This is because the calculation gives the area of a square, of given length.

For example, for a square of side 4 cm, area
= length × width
= 4 × 4
= $16\,\text{cm}^2$
A shorter way to write 4 × 4 is 4^2 that is, '4 squared'.

In 4^2, or 4^3, or 4^4, or 4^5 … the '2, 3, 4, 5 …' are known as **powers**, or **indices**. Sometimes they may be called **exponents**.
This number tells you how many times to multiply the number by itself.

3^2 = 3 to the power 2 or 3 squared = 3 × 3 = 9
5^3 = 5 to the power 3 or 5 cubed = 5 × 5 × 5 = 125
6^4 = 6 to the power 4 = 6 × 6 × 6 × 6 = 1296
2^8 = 2 to the power 8 = 2 × 2 × 2 × 2 × 2 × 2 × 2 × 2 = 256
8^1 = 8 to the power 1 = 8 = 8

Classroom challenge

1 Copy and complete these calculations.

13^2	13 × 13	169
14^2		
15^2		
16^2		
17^2		
18^2		
19^2		
20^2	20 × 20	400

2 Copy and complete this table so that you have a list of the first 20 square numbers.

Number	Square
1	1
2	4
3	9
4	16
5	25

You can download a copy of this table from the Bright Red website.

3 $3^4 = 3 \times 3 \times 3 \times 3 = 81$
Set these out in the same way.
a 3^5 c 4^3 e 5^3 g 7^4 i 6^6 k 10^4 m 1^7 o 12^1
b 11^2 d 2^6 f 3^3 h 11^3 j 10^6 l 4^4 n 5^1 p 8^3

4 $27 = 3 \times 3 \times 3 = 3^3$ (3 cubed)
Express the following numbers as cubes.
a $8 = \times \times = 2^3$ b 64 c 125 d 1000 e 1 000 000

STRETCH YOURSELF

5

A chess board full of rice! A mathematician's tale!

A story tells of a ruler of India who was so pleased when one of his staff developed the chessboard that he said the man could ask for anything he wanted.
This wise man also happened to be a mathematician, and asked for:
'one grain of rice on the first square of the chessboard'
'two grains of rice on the second square'
'four grains of rice on the third square'
'eight grains of rice on the fourth square'
and doubled up for each square on the chessboard.
The ruler thought this was a very modest request – and ordered that the rice be brought.
However, the rice quickly covered the chessboard, then filled the palace, then covered the region then … …

Can you work out how many grains of rice were on the last square?
Warning! Expect a big number.
And that was the last square only!
In fact, if you managed to collect all the rice, it would come to about 280 tonnes of rice!

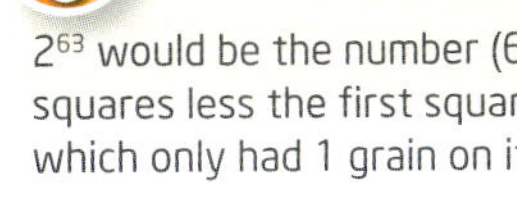

DON'T FORGET

2^{63} would be the number (64 squares less the first square which only had 1 grain on it)

Investigation

Try to continue this pattern. What do you notice?

$1^2 = 1 \times 1 = 1$
$11^2 = 11 \times 11 = 121$
$111^2 = 111 \times 111 = 12\,321$
$1111^2 = 1111 \times 1111 = 1\,234\,321$
…
$111\,111\,111^2$

In mathematics we often have 'opposite', 'reverse', or 'inverse' operations. That is, a way of getting back to the start.

How does that work?

If we ADD 3 to a number, we can SUBTRACT 3 to get back to the start.
If we MULTIPLY by 3 we can DIVIDE by 3 to get back to the start.
Add and subtract are inverse operations.
Multiply and divide are inverse operations.
Can we find an inverse operation to raising to a power?

SOLUTION

We know that $4^2 = 4 \times 4 = 16$
That is, 4 squared = 4 times itself = 16
Working backwards:
What number, times itself, gives 16? Answer 4
What number, times itself gives 9? We know that $3 \times 3 = 9$. Answer 3
What number, times itself gives 25? We know that $5 \times 5 = 25$. Answer 5

The inverse to **squaring a number** is called taking the **square root**.
The sign used is $\sqrt{\ }$.
Above, we could write:
What is $\sqrt{16}$? Answer 4 What is $\sqrt{9}$? Answer 3 What is $\sqrt{25}$? Answer 5

We will not look at other 'roots' in this chapter but you should know that they exist.
If you cube a number, to get back to start you would take the cube root, written $\sqrt[3]{\ }$
If you raise to the power 4, to get back to the start you would take the 4th root, written $\sqrt[4]{\ }$
For example,
$2^3 = 2 \times 2 \times 2 = 8$ therefore $\sqrt[3]{8} = 2$
$10^4 = 10 \times 10 \times 10 \times 10 = 10\,000$ therefore $\sqrt[4]{10\,000} = 10$

Classroom challenge

1 Copy and complete this table.

Number	Square root	Number	Square root
1		49	
4		64	
9		81	
16		100	
25		121	
36		144	

DON'T FORGET

Look back at the start of this chapter.
What do you do to find the area of a square?
Now reverse the process.

2 The area of this square field is $400\,m^2$. What is the length of the field?

3 The area of this game board is $900\,cm^2$. What is the length of its side?

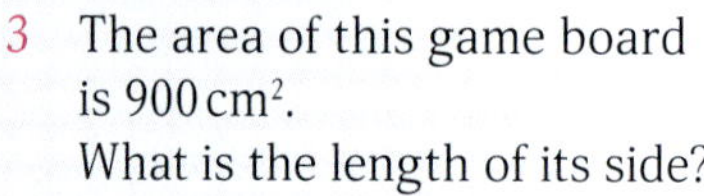

4 The area of this square stamp is $225\,mm^2$. What is the length of its side?

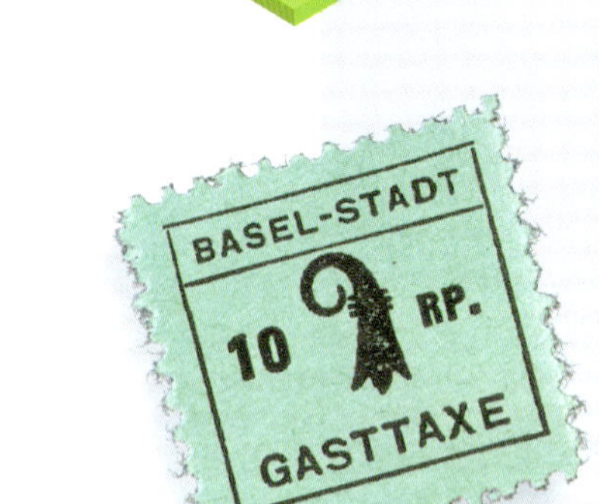

5 The area of this square photo frame is $625\,cm^2$. The length of the side of a square picture is 24 cm. Will it fit into the frame?

FRACTIONS, DECIMAL FRACTIONS AND PERCENTAGES

Calculations with fractions, decimals and percentages

I can solve problems by carrying out calculations with a wide range of fractions, decimal fractions and percentages, using my answers to make comparisons and informed choices for real-life situations. MNU 3-07a

What's coming up?

This Outcome and Experience will give you the opportunity to:

- convert fractions, decimal fractions or percentages into equivalent fractions, decimal fractions or percentages
- use your knowledge of fractions, decimal fractions and percentages to carry out calculations with and without a calculator.

What you already know

You have already learned how to:

- ✓ use your knowledge of equivalent forms of common fractions, decimal fractions and percentages
  - for example, $\frac{3}{4}$ = 0·75 = 75%, to solve problems
- ✓ calculate simple percentages of a quantity, and use this knowledge to solve problems in everyday contexts
  - for example, calculate the sale price of an item with a discount of 15%
- ✓ calculate simple fractions of a quantity and use this knowledge to solve problems
  - for example, find $\frac{3}{5}$ of 60
- ✓ identify multiples and factors of whole numbers and apply this knowledge and understanding of these when solving relevant problems in number, money and measurement.

Convert between fractions, decimals and percentages

Sometimes it is useful to choose which of fractions, decimals or percentages to use.
They all 'do the same thing' – but sometimes it is better to use one rather than the others.
A bit like trainers, school shoes and wellies! They all do the same thing – keep your feet safe and protected, but you would choose a different type of footwear depending on situation.

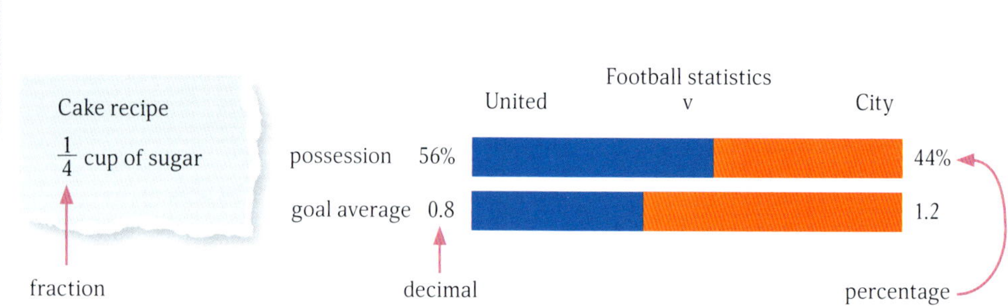

You would be unlikely to see a recipe which says, 'use 25% of a cup of sugar'.
You already know how to convert between common fractions, decimal fractions and percentages.
For example,

$50\% = \frac{50}{100} = 0{\cdot}5 = \frac{1}{2}$

$25\% = \frac{25}{100} = 0{\cdot}25 = \frac{1}{4}$

How does that work?

The same method works for percentages, decimal fractions and fractions which are not so 'common'.

EXAMPLE

Convert 35% to

a a decimal fraction

b a fraction.

SOLUTION

a $35\% = \frac{35}{100} = 35 \div 100 = 0{\cdot}35$

Hint: Remember that per cent means 'out of 100', that is, divide by 100. To divide by 100, the digits move 2 places right.

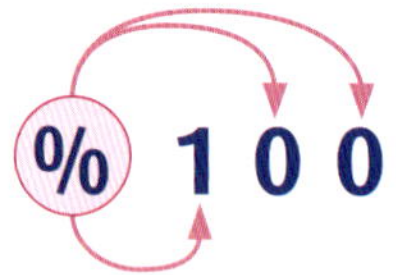

The per cent sign, %, should remind you of '100'. It is made up of a 1 and 2 zeros!

b $35\% = \frac{35}{100}$ Can this fraction be simplified?
Think back to the factors section.
What is the highest common factor of 35 and 100?
÷ 5

$= \frac{35 \div 5}{100 \div 5} = \frac{7}{20}$

EXAMPLE

Convert $\frac{3}{5}$ to:

a a percentage

b a decimal fraction.

SOLUTION

a $\frac{3}{5} \longrightarrow = \frac{3}{5} \times 100 = 300 \div 5 = 60\%$

b $\frac{3}{5} = 3 \div 5 = 0{\cdot}6$

Remember: to convert a fraction to a percentage, multiply by 100.

EXAMPLE

Convert 0·42 to:

a a percentage

b a decimal fraction.

SOLUTION

a $0{\cdot}42 \longrightarrow 0{\cdot}42 \times 100 = 42\%$

b $0{\cdot}42 = \frac{42}{100} \begin{smallmatrix}\div 2\\ \div 2\end{smallmatrix} = \frac{21}{50}$

Classroom challenge

A reminder of some common fractions and percentages.

1 Copy and complete this table to convert between common percentages, decimals and fractions.

Percentage	Decimal fraction	Fraction
50%		
	0·25	
		$\frac{3}{4}$
		$\frac{1}{3}$
$66\frac{2}{3}\%$		
	0·1	
		$\frac{1}{5}$
5%		

2 Copy and complete this table to convert between fractions, decimal fractions and percentages.

Percentage	Decimal fraction	Fraction
28%		
	0·3	
		$\frac{2}{5}$
	0·72	
6%		
		$\frac{17}{20}$
	0·08	
		$\frac{3}{8}$
72%		
	0·8	
45%		
		$\frac{5}{8}$
55%		
	0·01	
		$\frac{18}{200}$

3 Write these percentages as decimals.

a 42%
b 37%
c 23%
d 99%
e 5%
f 8%
g 1%
h 74%

4 Write these decimals as percentages.

a 0·78	c 0·92	e 0·03	g 0·07
b 0·26	d 0·44	f 0·09	h 0·125

5 Write these percentages as fractions. Simplify your fractions as far as possible.

a 50%	c 22%	e 15%	g 9%
b 70%	d 65%	f 4%	h 100%

6 Write these fractions as percentages.

a $\frac{36}{100}$	d $\frac{17}{50}$	g $\frac{7}{25}$	j $\frac{7}{10}$
b $\frac{72}{100}$	e $\frac{3}{20}$	h $\frac{14}{25}$	k $\frac{3}{4}$
c $\frac{24}{50}$	f $\frac{19}{20}$	i $\frac{4}{5}$	l $\frac{1}{10}$

7 Which one, of each pair, is smaller?

a 40% or $\frac{1}{4}$ b 0·4 or $\frac{3}{5}$ c 0·5 or 5%

8 Write these in ascending order – that is, starting with the smallest and going to largest.

12%, $\frac{3}{8}$, 0·0123, 42%, $\frac{1}{5}$, 0·45

9 Write these in descending order – that is, starting with largest and going to smallest. If necessary, round to 2 decimal places.

30%, $\frac{1}{3}$, 0·35, $\frac{3}{7}$, 45%, 0·04

10 There are 200 students in a school hall having lunch. Of these students, 124 have chosen yoghurt as part of their lunch.

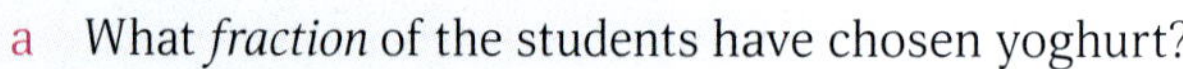

a What *fraction* of the students have chosen yoghurt?
b What *percentage* of the students have chosen yoghurt?
c What *decimal fraction* of the students have chosen yoghurt?

11 In a survey, $\frac{9}{10}$ of the students in a year group said their favourite subject was mathematics.
What *percentage* **did not say** their favourite subject was mathematics?

12 In a 2nd-year class, $\frac{3}{4}$ of the students said they could swim more than 300 metres $\frac{1}{10}$ of the class said they could only swim 100 metres.
a What percentage of the class can swim more than 300 metres?
b What percentage of the class can only swim up to 100 metres?
c What *fraction* of the class did not say they could swim these distances?

13 At a restaurant, customers could choose from three starters: breaded mushrooms, deep fried brie or scampi.
50% chose the scampi, and 24% chose the deep fried brie.
a What percentage chose breaded mushroom?
b What fraction chose breaded mushrooms?

14 In a bag of sweets, 40% are yellow and $\frac{7}{20}$ are blue.
a What percentage are blue?
b What percentage are neither yellow nor blue?
c What fraction are yellow?
d What fraction are yellow or blue?
e What fraction are neither yellow nor blue?

Calculate using fractions, decimals and percentages

When doing calculations, it is sometimes useful to use a format other than that given in the question.
You should also check to see if the question involves a 'common percentage'.

How does that work?

For example, if you had to work out 25% of £840, you would probably do something like this:

25% of £840

Which is the same as $\frac{1}{4} \times £840 = £210$

In this case, using a fraction made the calculation easier.

EXAMPLE

There are 12 000 people at a rugby match.
54% leave at half time.
How many stayed to watch the second half?

SOLUTION

54% of 12 000
= 54 ÷ 100 × 12 000

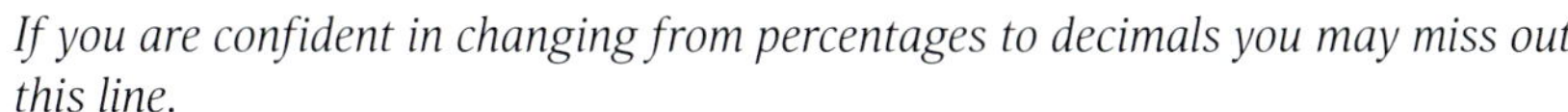

If you are confident in changing from percentages to decimals you may miss out this line.

= 0·54 × 12000
= 6400
Remaining for second half
= 12 000 – 6400
= 5600

Another method would be to calculate 46% of 12 000 ... can you explain why?

EXAMPLE

Grant has 224 football stickers. He gives away $\frac{3}{8}$ of them.
How many does he give away?

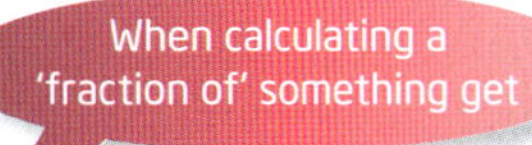

Multiply (×) and Divide (÷)

SOLUTION

$\frac{3}{8} \times 224$

= 224 × 3 ÷ 8 M a D Multiply by top number (numerator)
Divide by bottom number (denominator)

= 84
Grant gave away 84 football cards.

Classroom challenge

1 Calculate:

a 50% of £300
b 15% of 40 kg
c 40% of 600 metres
d 5% of £240
e 60% of 500 cm
f 2% of 250 g

2 Calculate:

a 26% of £370
b 43% of 850 kg
c 82% of 1 250 metres
d 17·5% of £80
e 0·5% of 30 kg
f 3·25% of £2500

3 There are 15 students in Mr Wilson's biology class. 80% of them passed a recent biology test.

a How many students passed the test?
b What fraction of the class did not pass the test?

4 Caroline and John are looking at new and used cars. They saw a new car for £15 000. They know that a used car typically sells for 85% of the cost of a new car. What price would they expect a used car to be? Round your answer to the nearest hundred pounds.

5 A large lecture theatre is being built at a university. There are 24 painters employed. Of these, 29% are employed to paint the interior. How many painters are employed to paint the interior? Round your answer to the nearest whole number.

6 The £1 coin weighs 9·5 g. It is made from 70% copper, 5% nickel and 25% zinc.

a How many grams of copper are there in the £1 coin?
b What fraction of the coin is nickel?
c What decimal fraction of the coin is zinc?

7 The picture shows a 50-pence coin. The coin weighs 8·0 g. It is made of cupronickel, which is 75% copper and the remainder nickel.

a What weight of the coin is copper?
b What fraction of the coin is nickel?
c The 50p is legal tender for amounts up to £10. What weight of nickel would be in £10 worth of fifty-pence pieces?

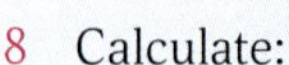

8 Calculate:

a $\frac{1}{3}$ of £24
b $\frac{1}{4}$ of 44 km
c $\frac{1}{9}$ of 36 kg
d $\frac{1}{8}$ of £64
e $\frac{1}{12}$ of 360 cm
f $\frac{1}{5}$ of £3000

9 Calculate:

a $\frac{3}{4}$ of £24
b $\frac{4}{5}$ of 200 km
c $\frac{3}{7}$ of 140 students
d $\frac{2}{9}$ of £540
e $\frac{5}{6}$ of 720 sweets
f $\frac{7}{8}$ of £4800

10 In a test there are 60 marks available.
Nadine achieves $\frac{4}{5}$ of the marks.
How many marks did Nadine achieve?

11 There are 160 students in 1st Year at Kant Academy.
$\frac{3}{8}$ of them travel by bus to school.
How many 1st year students travel by bus to school?

12 There are 36 houses in a street.
$\frac{11}{12}$ of them have satellite TV.
How many houses have satellite TV?

13 Grant has 360 stamps in his collection.
$\frac{3}{5}$ of them are foreign.
- a How many stamps are foreign?
- b What percentage are not foreign?

14 Bill and Ben sell some of their old games at a car boot sale.
They make £48.
Bill gets $\frac{5}{8}$ of the money.
- a How much money will Bill get?
- b What percentage of the money will Ben get?

15 A Pets 4 U shop has 30 animals. $\frac{3}{5}$ of them are puppies.
A Pets R Us shop has 21 animals. $\frac{5}{7}$ of them are puppies.
Which shop has more puppies?

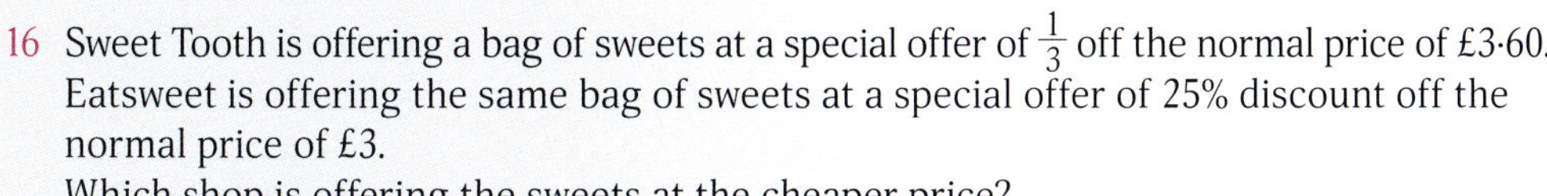

16 Sweet Tooth is offering a bag of sweets at a special offer of $\frac{1}{3}$ off the normal price of £3·60.
Eatsweet is offering the same bag of sweets at a special offer of 25% discount off the normal price of £3.
Which shop is offering the sweets at the cheaper price?

17 There are 600 students in a school.
- a If lunch has to be prepared for $\frac{2}{3}$ of the students, how many lunches are required?
- b How many more lunches would be required if $\frac{3}{4}$ of the students wanted lunch?

18 A factory carries out a survey of its 500 workers.
$\frac{2}{5}$ say they would like more overtime.
80% said they were happy with the holiday pattern.
0·3 of them said they would like a shorter lunch break and to finish earlier.
- a How many said they would like more overtime?
- b How many said they were happy with the holiday pattern?
- c How many would like a shorter lunch break and an earlier finishing time?

19 Mark scored 40 out of 50 in a test. What percentage is this?

20 A check of light bulbs showed that 15 out of 200 were faulty.
What percentage were faulty?

21 4 kilometres of a 20 kilometre road had road works on it.
What percentage of the road had roadworks?

22 Joe scored $\frac{15}{20}$ in a test. Jack scored $\frac{35}{50}$ in another test.
Who had the better percentage mark?

23 12 out of 15 grams of a coin is copper.
What percentage is this?

24 In a game of chance, Penny had a probability of $\frac{5}{8}$ of winning.
The sign on the game said '70% chance of winning'.
Is this claim valid?

25 Zudain scored 15 goals in 40 games.
What is his percentage scoring rate?

Multiplying fractions

Sometimes calculations involving fractions include multiplying fractions together.

How does that work?

There are 3 steps to complete when multiplying fractions.

1 Multiply across the top numbers (numerators).
2 Multiply across the bottom numbers (denominators).
3 Simplify if possible.

EXAMPLE

$\frac{1}{2} \times \frac{2}{5}$

1 Multiply across top:
2 Multiply across bottom: $\frac{1}{2} \times \frac{2}{5}$ $\quad \frac{1 \times 2}{2 \times 5} = \frac{2}{10}$
3 Simplify: $\frac{2}{10} = \frac{1}{5}$ We have 'cancelled down' by 2.

This method works with whole numbers as well.

EXAMPLE

What is $\frac{3}{5} \times 4$? 4 can be written as $\frac{4}{1}$

$\frac{3}{5} \times 4 = \frac{3}{5} \times \frac{4}{1} = \frac{3 \times 4}{5 \times 1} = \frac{12}{5}$

Classroom challenge

1 Calculate, simplifying your answer where possible.

a $\frac{1}{5} \times \frac{2}{5}$
b $\frac{1}{2} \times \frac{3}{4}$
c $\frac{1}{4} \times \frac{2}{5}$
d $\frac{1}{6} \times \frac{3}{4}$
e $\frac{2}{7} \times \frac{3}{5}$
f $\frac{3}{8} \times \frac{1}{7}$
g $\frac{3}{10} \times \frac{7}{10}$
h $\frac{4}{9} \times \frac{2}{3}$
i $\frac{4}{11} \times \frac{3}{5}$
j $\frac{3}{5} \times 6$
k $\frac{3}{7} \times 3$
l $5 \times \frac{2}{3}$
m $\frac{3}{8} \times \frac{2}{7}$
n $\frac{3}{5} \times \frac{5}{6}$
o $\frac{4}{9} \times \frac{3}{7}$
p $\frac{5}{8} \times \frac{2}{3}$
q $\frac{2}{9} \times \frac{3}{8}$
r $\frac{7}{10} \times \frac{9}{10}$

An alternative method is to simplify or cancel first, if you can, and then multiply.

The advantage with this method is that you would be multiplying smaller numbers.

How does that work?

What is $\frac{4}{9} \times \frac{15}{16}$?

Step 1 $\frac{4}{9} \times \frac{15}{16}$

Step 2 Notice that the 4 and the 16 (one top and one bottom can cancel).
4 is the highest common factor of 4 and 16.

Giving $\frac{1}{9} \times \frac{15}{4}$

Step 3 Check whether further simplifying can take place.
3 is the highest common factor of 15 and 9.

Giving $\frac{1}{3} \times \frac{5}{4}$

Step 4 Multiply: $\frac{5}{12}$

Classroom challenge

1 Use the method above to answer these questions.

a $\frac{8}{9} \times \frac{15}{16}$

b $\frac{4}{9} \times \frac{15}{28}$

c $\frac{7}{12} \times \frac{8}{11}$

d $\frac{7}{12} \times \frac{8}{21}$

e $\frac{7}{11} \times \frac{22}{35}$

f $\frac{5}{14} \times \frac{21}{30}$

g $\frac{2}{7} \times \frac{35}{36}$

h $\frac{3}{17} \times \frac{51}{60}$

i $\frac{8}{11} \times \frac{33}{80}$

Head to www.brightredpublishing.co.uk and download our Teaching Notes to find:
- an equivalent fractions poster
- a 'fraction strips' poster.

Convert between mixed numbers and fractions

“Having used practical, pictorial and written methods to develop my understanding, I can convert between whole or mixed numbers and fractions.” MTH 3-07c

What's coming up?

This Outcome and Experience will give you the opportunity to:

- convert between whole or mixed numbers, improper fractions and decimal fractions.

What you already know

You have already learned how to:

- create equivalent fractions and use this knowledge to put a set of most commonly used fractions in order
- express fractions in their simplest form.

So far, we have worked with what are called ‘proper fractions’.

For example, $\frac{3}{4}$, or $\frac{2}{7}$, or $\frac{5}{12}$.

Here the numerator is smaller than the denominator.

However, we sometimes come across fractions like $\frac{4}{3}$, or $\frac{8}{3}$ or $\frac{15}{11}$

Here the numerator is greater than the denominator.

These are called ‘improper fractions’.

How does that work?

EXAMPLE

Show that the numbers $3\frac{1}{4}$ and $\frac{13}{4}$ are equivalent.

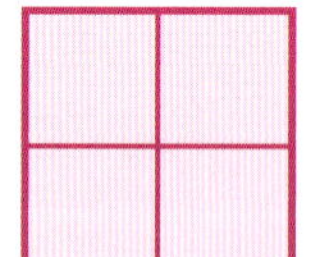
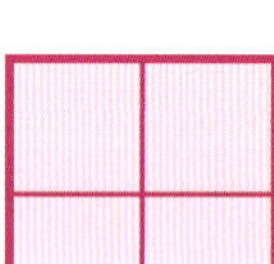
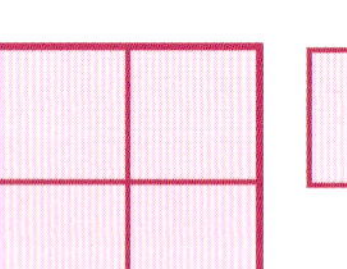

SOLUTION

Use a diagram:
You can think of 3 ¼ as 3 ‘wholes’ and 1 quarter.
You can see that if you count the ‘quarters’

there are	4 + 4 + 4 + 1	= 13 quarters	$= \frac{13}{4}$
That is	3 lots of 4 plus the 1		$= \frac{13}{4}$
or	$3 \times 4 + 1$		$= \frac{13}{4}$

$3\frac{1}{4}$ is called a **mixed number** as it is a mixture of a whole number and a fraction.

When changing from a mixed number to an improper fraction, like before, you should get:

Look at the example above.

Multiply (×) Add (+) Done

$3\frac{1}{4}$ Multiply the 3 and 4 (how many $\frac{1}{4}$s in 3 wholes) then add the 1 quarter.

$3\frac{1}{4}$ (+, ×) $= 4 \times 3\ (=12) + 1 = 13$ giving $\frac{13}{4}$

EXAMPLE

$5\frac{3}{7}$ $= 7 \times 5\ (=35) + 3 = 38$ giving $\frac{38}{7}$

Working the other way, what would $\frac{38}{7}$ be as a mixed number?

SOLUTION

$\frac{38}{7}$ is a dividing sum.

$38 \div 7 = 5$ remainder 3 that is, 5 and 3 sevenths left over giving $5\frac{3}{7}$

EXAMPLE

$\frac{22}{5}$ as a mixed number is:

$22 \div 5 = 4$ remainder 2 that is, 4 and 2 fifths left over giving $4\frac{2}{5}$

Classroom challenge

1 Write the numbers represented by these diagrams in two different ways. The first one has been done for you.

a 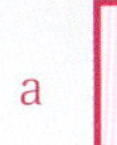  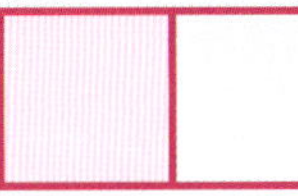 shaded = 3 halves $= \frac{3}{2} = 1\frac{1}{2}$

b 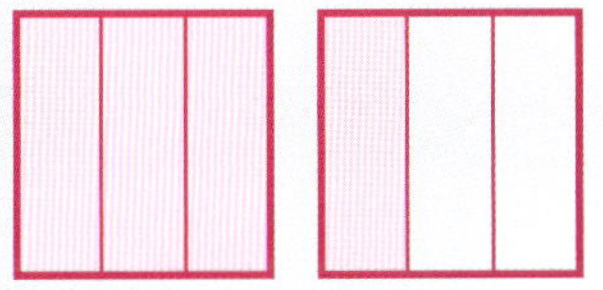

c

d

e

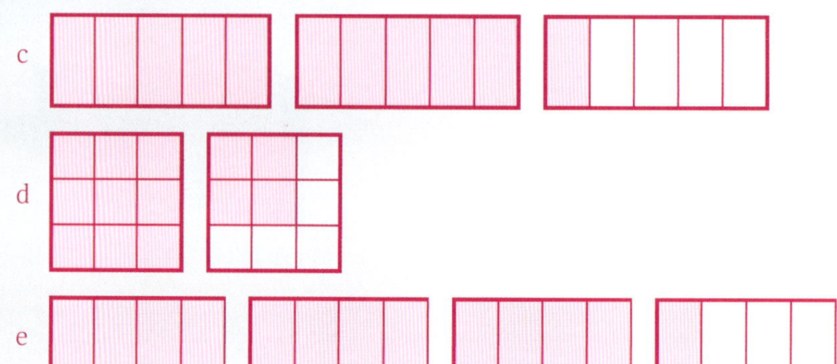

2 Draw diagrams to represent these mixed numbers. Write each of the mixed numbers below as improper fractions.

a $2\frac{3}{5}$ b $1\frac{3}{8}$ c $3\frac{2}{3}$ d $1\frac{7}{10}$

3 Draw diagrams to represent these improper fractions. Write each of the improper fractions below as a mixed number.

a $\frac{11}{5}$ b $\frac{13}{8}$ c $\frac{14}{3}$ d $\frac{11}{10}$

4 Write these improper fractions as mixed numbers.

a $\frac{7}{2}$ e $\frac{8}{5}$ i $\frac{57}{9}$ m $\frac{25}{12}$

b $\frac{5}{3}$ f $\frac{12}{5}$ j $\frac{39}{9}$ n $\frac{37}{11}$

c $\frac{7}{3}$ g $\frac{15}{7}$ k $\frac{31}{10}$ o $\frac{54}{11}$

d $\frac{7}{4}$ h $\frac{30}{7}$ l $\frac{103}{10}$ p $\frac{39}{12}$

5 Write these mixed numbers as improper fractions. Remember: get 'MAd'.

a $1\frac{1}{2}$ e $5\frac{1}{2}$ i $8\frac{1}{4}$ m $4\frac{3}{10}$

b $2\frac{1}{3}$ f $3\frac{2}{3}$ j $4\frac{3}{7}$ n $3\frac{4}{11}$

c $1\frac{4}{5}$ g $6\frac{3}{4}$ k $3\frac{7}{8}$ o $2\frac{1}{12}$

d $3\frac{1}{4}$ h $4\frac{2}{5}$ l $7\frac{3}{5}$ p $4\frac{7}{9}$

6 Write these fractions in ascending order, that is, from smallest to biggest.

$\frac{18}{5}$, $6\frac{1}{2}$, $4\frac{1}{4}$, $7\frac{1}{3}$, $\frac{16}{3}$

7 Write these fractions in descending order, that is, from biggest to smallest.

$2\frac{3}{5}$, $\frac{13}{2}$, $4\frac{1}{2}$, $\frac{23}{5}$, $7\frac{1}{4}$

8 Copy each pair of fractions.
Insert either > greater than < less than or = equal to between each pair to make it a true sentence.

a $2\frac{3}{5}$ $\frac{13}{5}$ c $4\frac{1}{4}$ $\frac{11}{2}$ e $\frac{17}{5}$ $3\frac{2}{5}$

b $3\frac{3}{5}$ $\frac{17}{5}$ d $\frac{14}{3}$ $4\frac{1}{3}$ f $\frac{23}{7}$ $3\frac{1}{7}$

9 Explain why $\frac{38}{8}$ is equal to $4\frac{3}{4}$.

10 A puppy dog is 32 months old.
Write down the age of the puppy in years, as a mixed number.
Simplify your answer as far as possible.

Add and subtract fractions

By applying my knowledge of equivalent fractions and common multiples, I can add and subtract commonly used fractions.

MTH 3-07b

What's coming up?

This Outcome and Experience will give you the opportunity to:

- add and subtract whole numbers and fractions, including when changing a denominator.

What you already know

You have already learned how to:

- create equivalent fractions and use this knowledge to put a set of most commonly used fractions in order
- express fractions in their simplest form
- identify multiples and factors of whole numbers and apply your knowledge and understanding of these when solving relevant problems in number, money and measurement.

Adding and subtracting when the denominators are the same

How does that work?

EXAMPLE

$\frac{1}{5} + \frac{3}{5}$ Denominators are the same so '1' has been cut into the same size pieces.

$\frac{1}{5} + \frac{3}{5}$ When the denominators are the same, you simply add the numerators.

$\frac{1}{5} + \frac{3}{5} = \frac{1 + 3}{5} = \frac{4}{5}$ In the diagram above, 4 fifths have been shaded in total.

EXAMPLE

$\frac{5}{8} - \frac{3}{8}$ When the denominators are the same, you simply subtract the numerators.

$\frac{5}{8} - \frac{3}{8} = \frac{5}{8} - \frac{3}{8} = \frac{2}{8} = \frac{1}{4}$ Always simplify a fraction if possible.

When working with fractions, the denominators are not always the same.

To add or subtract fractions with different denominators, use the skills you learned when working with lowest common multiples.

How does that work?

EXAMPLE

$\frac{1}{4} + \frac{2}{3}$ Here you should determine the lowest common multiple of 3 and 4, which is 12.

Now change both fractions to twelfths.

$\frac{1 \times 3}{4 \times 3} + \frac{2 \times 4}{3 \times 4}$ Multiply 1st fraction by $\frac{3}{3}$ to give denominator of 12.

Multiply 2nd fraction by $\frac{4}{4}$ to give denominator of 12.

$\frac{3}{12} + \frac{8}{12} = \frac{11}{12}$

EXAMPLE

$\frac{15}{16} - \frac{3}{4}$ The **LCM** of 16 and 4 is 16.

$\frac{15 \times 1}{16 \times 1} - \frac{3 \times 4}{4 \times 4}$ Multiply 1st fraction by $\frac{1}{1}$ to give denominator of 16.

Multiply 2nd fraction by $\frac{4}{4}$ to give denominator of 16.

$\frac{15}{16} - \frac{12}{16} = \frac{3}{16}$

Classroom challenge

1 Calculate:

a $\frac{4}{7} + \frac{1}{7}$
b $\frac{4}{7} - \frac{1}{7}$
c $\frac{3}{8} + \frac{3}{8}$
d $\frac{1}{9} + \frac{7}{9}$
e $\frac{3}{10} + \frac{5}{10}$
f $\frac{9}{10} - \frac{1}{10}$
g $\frac{3}{11} + \frac{5}{11}$
h $\frac{7}{11} - \frac{1}{11}$
i $\frac{7}{20} + \frac{9}{20}$
j $\frac{6}{13} + \frac{7}{13}$
k $\frac{9}{13} - \frac{9}{13}$
l $\frac{23}{100} + \frac{51}{100}$

2 Calculate these. The first one has been done for you.

a $\frac{1}{2} + \frac{1}{5}$ Remember: find the LCM of 2 and 5 (10).

$\frac{1}{2} \times \frac{5}{5} = \frac{5}{10}$ $\frac{1}{5} \times \frac{2}{2} = \frac{2}{10}$

$\frac{1}{2} + \frac{1}{5} = \frac{5}{10} + \frac{2}{10} = \frac{7}{10}$

b $\frac{1}{6} + \frac{2}{3}$

c $\frac{4}{5} + \frac{2}{3}$

d $\frac{4}{7} - \frac{1}{3}$

e $\frac{5}{6} - \frac{2}{3}$

f $\frac{3}{4} + \frac{2}{3}$

g $\frac{1}{4} + \frac{1}{5}$

h $\frac{2}{3} + \frac{3}{5}$

i $\frac{5}{6} - \frac{1}{2}$

j $\frac{7}{8} - \frac{3}{4}$

k $\frac{6}{7} - \frac{2}{3}$

l $\frac{4}{5} - \frac{3}{4}$

m $\frac{2}{3} + \frac{1}{7}$

n $\frac{9}{10} - \frac{1}{3}$

o $\frac{3}{16} + \frac{1}{4}$

p $\frac{15}{16} - \frac{7}{8}$

3 Pat's birthday cake is cut into 8 equal slices. Mike eats $\frac{3}{8}$ of the cake and Calum eats $\frac{1}{4}$ of the cake.

a What fraction of Pat's birthday cake is left?

b How many slices are left?

4 Grigor runs $\frac{3}{4}$ of a kilometre. He then runs a further $\frac{1}{5}$ of a kilometre.

a What fraction of a kilometre has he now run?

b How much further will he have to run to complete a distance of 2 kilometres?

5 Maria saves some programs onto a memory stick. She uses $\frac{2}{5}$ of the space. Scott saves his programs to the same memory stick. He uses $\frac{1}{3}$ of the memory stick's space.

a What fraction of the memory stick is now used?

b What fraction of the memory stick is unused?

c Maria deletes a program which used $\frac{1}{15}$ of the space on the memory stick. What fraction of space is left now?

When adding or subtracting mixed numbers, it is often better to change the mixed number to an improper fraction first, and then do the calculation.

How does that work?

EXAMPLE

Calculate $3\frac{1}{4} - 1\frac{2}{5}$

SOLUTION

$= \frac{13}{4} - \frac{7}{5}$ The LCM of 4 and 5 is 20.

$= \frac{65}{20} - \frac{28}{20}$

$= \frac{37}{20}$

$= 1\frac{17}{20}$

Classroom challenge

1 Calculate:

a $1\frac{1}{2} + 1\frac{1}{3}$
b $2\frac{1}{2} + 1\frac{3}{4}$
c $3\frac{1}{2} - 1\frac{1}{3}$
d $1\frac{2}{7} + 2\frac{3}{14}$
e $3\frac{1}{4} - 1\frac{3}{5}$
f $3\frac{1}{2} - 1\frac{3}{4}$
g $4\frac{2}{3} - 2\frac{1}{2}$
h $6\frac{1}{4} - 2\frac{2}{5}$
i $2\frac{1}{2} + 1\frac{5}{8}$
j $3\frac{2}{3} + 1\frac{5}{9}$
k $2\frac{1}{10} + 3\frac{2}{5}$
l $4\frac{5}{12} - 1\frac{5}{6}$

2 Sammi measures the height of his sunflower and finds it is $1\frac{3}{5}$ metres tall. Three weeks later she measures it again and finds that it is now $2\frac{1}{4}$ metres tall. How much did the sunflower grow during those three weeks?

3 Graeme's water barrel, in his garden, is $\frac{3}{4}$ full. During a shower of rain, $\frac{1}{6}$ of the barrel more water is added to it. How full is Graeme's water barrel now?

4 Rosie drank $1\frac{1}{4}$ cups of milk at breakfast and $2\frac{1}{3}$ cups of milk at dinner. How many cups of milk did Rosie drink altogether?

5 Louise bought $1\frac{2}{5}$ kg of beef mince and $2\frac{1}{4}$ kg of pork mince from the butcher. How much mince did Louise buy in total?

6 Jennifer made $11\frac{2}{3}$ cups of lemonade. She drank $6\frac{3}{4}$ cups of lemonade. How many cups of lemonade did she have left?

7 A café had $8\frac{1}{2}$ kg of coffee beans when they opened on Monday morning. By the end of Wednesday, they had used $6\frac{3}{4}$ kg of the coffee beans. What weight of coffee beans did they have left?

8 Jack ran $2\frac{1}{3}$ km. Jill ran $3\frac{2}{5}$ km. How much farther did Jill run compared to Jack?

9 Julie and Jeremy went fishing. Julie caught $3\frac{2}{5}$ kg of fish. Jeremy caught $2\frac{1}{4}$ kg of fish. Calculate the total weight of the fish caught by Julie and Jeremy.

10 Jake has a wooden beam which is $5\frac{3}{4}$ metres long. He cuts off a section $1\frac{2}{3}$ metres long. He then cuts off another section, $1\frac{1}{4}$ metres long. What length of wooden beam does Jake have left?

STRETCH YOURSELF

11 A fuel tank has $42\frac{3}{4}$ litres of fuel in it. $31\frac{5}{12}$ litres are used. The tank was then topped up with $17\frac{2}{3}$ litres of fuel. What is the final volume, in litres, of fuel in the tank?

12 Iain has $3\frac{1}{2}$ bottles of fresh orange juice in his fridge. He drank $\frac{4}{5}$ of a bottle in the morning and $1\frac{1}{4}$ bottles in the afternoon. How much fresh orange does Iain have left?

Proportion and ratio

I can show how quantities that are related can be increased or decreased proportionally and apply this to solve problems in everyday contexts. MNU 3-08a

What's coming up?

This Outcome and Experience will give you the opportunity to:
- solve problems in which related quantities are increased or decreased proportionally
- express quantities as a ratio and where appropriate simplify, for example, 'if there are 6 teachers and 60 children in a school, find the ratio of the number of teachers to the total amount of teachers and children'.

What you already know

You have already learned how to:
- calculate simple fractions of a quantity and use this knowledge to solve problems
  - for example, find $\frac{3}{5}$ of 60
- create equivalent fractions and use this knowledge to put a set of fractions in order
- express fractions in their simplest form
- identify multiples and factors of whole numbers and apply your knowledge and understanding of these when solving relevant problems in number, money and measurement.

Because ratio and proportion are often taught together, many people think that they are the same thing.

Really, however, they are different.

A ratio is a fraction that compares two quantities of the same 'unit'.

Keywords in ratio are 'for every'. For example, there are 3 red flowers **for every** 4 blue flowers

This is written as 3:4.

Proportion compares one quantity with the total.

Keywords in proportion are 'out of'. For example, there are 3 red flowers **out of** a total of 7 flowers

This would be written as $\frac{3}{7}$.

We will split the two topics into two separate exercises.

Proportion

How does that work?

Proportion compares a quantity with the total.

Keywords are **'out of'**.

EXAMPLE

What proportion of this metre stick is shaded?
3 pieces are shaded **out of** a total of 5 pieces.
Therefore, shaded part is $\frac{3}{5}$

EXAMPLE

A fruit juice is made by mixing 30 ml of blackcurrant concentrate with 70 ml of water.
What proportion of the juice is blackcurrant concentrate?
30 ml out of a total of 100 ml is blackcurrant concentrate.
Therefore, the proportion is $\frac{30}{100} = \frac{3}{10}$.

Proportion can be used to calculate quantities that are related.

EXAMPLE

5 pencils cost £3.
How much would 8 pencils cost?

SOLUTION

A good method is to set up a table and work out the cost of 1 pencil.
Then you can calculate the cost of any number of pencils.

	Pencils	Cost £	
	5	3	
÷ 5	1	0·60	÷ 5
× 8	8	£4·80	× 8

The cost of 8 pencils is £4·80.

Classroom challenge

1 For each of these metre sticks, write down the proportion that is:

i shaded

ii unshaded.

a

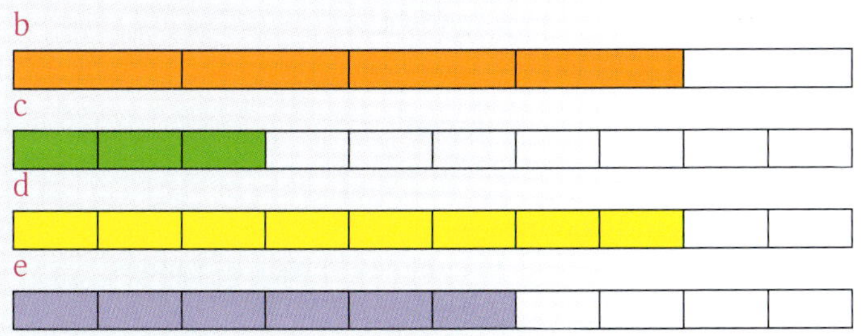

b

c

d

e

2 For each of these boxes of marbles, write down the proportion of:

i red marbles

ii green marbles.

a

b

c

d

3 The table shows how much raspberry concentrate and water is used to mix fruit juices.

Copy the table and complete the column to show the proportion of raspberry concentrate.

Raspberry concentrate (ml)	Water (ml)	Total volume (ml)	Proportion of raspberry concentrate
10	40	50	$\frac{1}{5}$
20	70		
45	55		
60	140		
15	35		

DON'T FORGET

Remember to set out your working in a table as in the example.

4 6 pencils cost £1·80. How much will 10 of the same pencils cost?

5 8 slices of bacon cost £3·20. What will 5 slices of bacon cost?

6 50 apples cost £125. What would 75 apples cost?

7 Karen reads 4 pages of a book in 18 minutes.
How long would Karen take, at the same pace, to read 30 pages?

8 A car, travelling at a constant speed, covers 120 miles in 3 hours.
How long would it take to cover 200 miles?

9 David makes hand puppets.
He can make 15 hand puppets in 2 hours.
How many hand puppets could David make in a working day of 8 hours?

10 A sausage machine produces 250 sausages in 5 minutes.
How long would it take to fulfil an order for 1000 sausages?

11 A pastry chef makes 20 Danish pastries in 35 minutes.
How long would it take her to make 50 Danish pastries?

12 Nina completed a 30 mile hill-walk in 10 hours.
At the same pace, how long would it take Nina to complete a walk of 9 miles?

13 A bookshop sells 16 books in 5 days.
At the same rate, how long will it take to sell 112 books?

14 Steve washed 15 cars in 3 hours.
At the same rate, how long will it take Steve to wash 7 cars?

15 John, a photographer, takes 15 pictures in 5 minutes.
How long will it take him to take 120 pictures?

16 Two water towers hold 420 gallons of water.
How many gallons of water would 5 water towers hold?

17 A car factory produces 50 cars in 6 days.
How many days would it take to complete an order for 800 cars?

18 A crane can load 82 container lorries in 16 hours.
How long would it take to load 287 container lorries?

19 A pile of paper has 150 sheets in it. The pile is 2·6 centimetres high.
How high would a pile be if it contained 120 sheets?

20 28 bottles of olive oil costs £111·72.
What would 16 bottles of olive oil cost?

Ratio

How does that work?

EXAMPLE

Remember, ratio compares quantities of the same 'unit'.
Keywords are '**for every**'.

In this diagram of a metre stick, what is the ratio of shaded to unshaded parts?
There are 2 shaded parts **for every** 3 unshaded parts.

Therefore, the ratio of	shaded	to	unshaded
is	2	:	3

Note that the order is important!

The ratio of	unshaded	to	shaded
is	3	:	2

EXAMPLE

3 litres of red paint is mixed with 4 litres of blue paint.

a What is the ratio of red to blue paint?
b What is the ratio of blue to red paint?
c What is the ratio of red to total paint?

SOLUTION

a	There are 3 litres of red **for every** 4 litres of blue paint.	Ratio of red : blue is	3 : 4
b	There are 4 litres of blue **for every** 3 litres of red paint.	Ratio of blue : red is	4 : 3
c	There are 3 litres of red **for every** 7 litres of paint.	Ratio of red : total is	3 : 7

EXAMPLE

Ratios can be simplified, in a similar way to simplifying fractions.
As with fractions, looking for the highest common factor (HCF) will help you to simplify a ratio.

A ratio of 8 : 24 can be simplified to make it easier to work with
The HCF of 8 and 24 is 8.

÷8	8	:	24	÷8
	1	:	3	

This ratio will now be easier to work with.

Ratios always use whole numbers. If you get a ratio that has a fraction in it, multiply up to get whole numbers.

EXAMPLE

Simplify a ratio of $\frac{1}{5} : 3$ Multiply both sides by 5.

giving a ratio of 1 : 15

EXAMPLE

Using ratios to carry out calculations.

A scale on a map is 1 : 20 000.
What is the real distance of a route that measures 7 centimetres on the map?

SOLUTION

	Scale	:	Real	
×7	1	:	20 000	×7
	7	:	140 000 cm	
			= 1400 m	
			= 1·4 km	

EXAMPLE

In a maths class the ratio of boys to girls is 2 : 3.
There are 12 boys in the class.
How many girls are there in the class?

SOLUTION

Ratio	Boys	:	Girls	
	2	:	3	
×6	12		18	×6

There are 18 girls in the class.

Classroom challenge

1 Orange squash is mixed with water in the ratio 1 : 8.
Copy and complete the table to show the amounts of water required for the different amounts of orange squash.

Orange squash (ml)	Water (ml)
1	8
10	
50	
5	

2 Write each of these ratios in its simplest form.

a 4 : 8
b 3 : 9
c 5 : 25
d 2 : 10
e 6 : 2
f 21 : 7
g 48 : 16
h 54 : 9
i 14 : 21
j 35 : 15
k 18 : 24
l 15 : 10
m 33 : 77
n 100 : 60
o 25 : 40
p 144 : 48

3 Zeishan mixes 600 ml of orange juice with 900 ml of peach juice to make a fruit drink.
a Write the ratio of orange juice to peach juice in its simplest form.
b Write the ratio of orange juice to the total volume in its simplest form.

4 A builder mixes 8 shovels of cement with 24 shovels of sand.
Write the ratio of cement to sand, in its simplest form.

5 In a cake mix the ratio of butter to jam is 300 : 800.
Write this ratio in its simplest form.

6 In a school there are 650 students and 40 teachers.
a Write the ratio of students to teachers in its simplest form.
b Write the ratio of teachers to students in its simplest form.

7 The ratio of girls to boys in a mathematics class is 5 : 3.
If there are 20 girls in the class, how many boys are there?

8 The ratio of strawberries to blackcurrants in a pie is 2 : 7.
If there are 20 g of strawberries in the pie, what weight of blackcurrants is there?

9 The ratio of a model car to the real car is 3 : 250.
The model car is 4·5 cm long.
How long is the real car?

10 On a school trip the ratio of teachers : students is 1 : 10.
If there are 80 students going on the trip, how many teachers will be required?

11 A map is drawn to a scale of 1 : 50 000.
Calculate, in kilometres, what the real distance would be given these measurements on the map.
a 3 cm b 8 cm c 12 cm d 22 cm

12 A map is drawn to a scale of 1 : 250 000.
The distance between two towns is 80 kilometres.
How far apart are the towns on the map?

DON'T FORGET

You may wish to 'turn the ratio around'.

How does that work?

EXAMPLE

Sharing in a ratio

Matthew and Mark share the profit from their business in the ratio 3:5.
This means that for every £8 profit, Matthew gets £3 and Mark gets £5.
Do you see where the £8 comes from?
Yes, it is the 'total of parts'.
If the business made a profit of £16 000, how much would each get?

Ratio of 3:5	is a total of 8 parts (3 + 5 = 8)		
	A profit of £16 000 shared into 8 parts	=	£2000 for each part.
Matthew would get	3 parts at £2000 (3 × £2000)	=	£6000
Mark would get	5 parts at £2000 (5 × £2000)	=	£10 000
	Check that these total	=	£16 000

We can look at this using a diagram.
A ratio of 3:5

Matthew ○ ○ ○

8 circles → 8 parts

Mark ○ ○ ○ ○ ○

£16 000 ÷ 8 = £2000
Put £2000 into each circle.
and it is easy to see how much each person gets.

EXAMPLE

Mandy, Ricky and James share £240 in the ratio 2 : 3 : 1.
How much does each person get?

A ratio of	2 : 3 : 1	=	6 parts	(2 + 3 + 1 = 6)
£240 ÷ 6		=	£40 for each part	
Mandy gets	2 × £40	=	£80	
Ricky gets	3 × £40	=	£120	
James gets	1 × £40	=	£40	
Check total is correct		=	£240	

Classroom challenge

1 a Share £40 in the ratio 2:3
  b Share £35 in the ratio 2:5
  c Share 60 kg in the ratio 7:3
  d Share 80 cm in the ratio 3:1

2 a Share £60 in the ratio 1:2:3
  b Share £108 in the ratio 5:3:4
  c Share 120 kg in the ratio 6:2:4
  d Share 75 litres in the ratio 5:9:11

3 Luke and John share the £90 profit they made at a car boot sale in the ratio 4:5.
How much did each receive?

4 Blue and yellow paint were mixed in the ratio 4:7 to make green paint.
How much of each colour is required to make 132 ml of green paint?

5 In a chemistry experiment, water and acid are mixed in the ratio 8:1.
A bottle contains 135 ml of the mixture.
  a How much water is in the mixture?
  b How much acid is in the mixture?

6 Abigail, Brian and Charlotte share their birthday money of £360 in the same ratio as their ages 8:6:4.
How much money does each person get?

7 In a fruit drink, peach juice, pineapple juice and lemon juice are mixed in the ratio 3:5:1.
How much of each type of juice is required to make:
  a 90 ml of the drink
  b 270 ml of the drink
  c 450 ml of the drink?

8 In an Art class, yellow, blue and red paint are mixed together to make 300 ml of another colour.
How much of each colour paint would be needed if the paints were mixed in the ratio:
  a 1:2:3
  b 2:5:3
  c 3:7:5
  d 8:8:14?

FINANCIAL SKILLS

MONEY

When considering how to spend my money, I can source, compare and contrast different contracts and services, discuss their advantages and disadvantages, and explain which offer best value to me. MNU 3-09a

What's coming up?

This Outcome and Experience will give you the opportunity to:

- demonstrate understanding of best value in relation to contracts and services when comparing products
- choose the best value for a given situation and justify your choice.

What you already know

You have already learned how to:

- ✔ carry out money calculations involving the four operations
- ✔ compare costs and determine affordability within a given budget
- ✔ demonstrate understanding of the benefits and risks of using bank cards and digital technologies.

Best value

Calculate best value in deals and offers

When buying goods, you should always look for the best value.

One way to find better or best value is to compare prices for a 'unit'.

For example, you may compare prices per tin, or prices per litre, or prices per kilogram.

You should also look to see if there are special offers or discounts.

For example, 'buy 2 get one free' or 'buy 1 get 1 half-price' or 20% discount.

How does that work?

EXAMPLE

Toni's store sells tubs of 'Classic Couscous' at 5 for £1·80.
Lenny's shop sells the same tubs at 3 for £1·05.
Which shop offers the better value?

SOLUTION

Calculate the cost 'per tub', that is, the cost for 1 tub.

Toni's store		Lenny's shop	
Tubs	Cost (£)	Tubs	Cost (£)
5	1·80	3	1·05
1	1·80 ÷ 5	1	1·05 ÷ 3
=	£0·36 per tub	=	£0·35 per tub

Lenny's shop gives the better value.

You may wish to look back at your work on proportion.

EXAMPLE

A 2 litre bottle of 'Fizz' costs £2·70.
A 1·5 litre bottle of 'Fizz' costs £2·07.
Which is the better value?

SOLUTION

Litres	Cost (£)	Litres	Cost (£)
2	2·70	1·5	2·07
1	2·70 ÷ 2	1	2·07 ÷ 1·5
=	£1·35 per litre	=	£1·38 per litre

The 2 litre bottle represents the better value.

EXAMPLE

'Sports 4 Me' sell a tennis racquet for £120. They are offering a 10% discount.
'My Sport' sell the same tennis racquet for £128. They are offering a 15% discount.
Stella wants to buy the racquet.
Which shop would offer her the better value?

SOLUTION

Sports 4 Me		My Sport	
Original cost	£120	Original cost	£128·00
10% discount	– £12	15% discount	– £19·20
Offer price	£108	Offer price	£108·80

Sports 4 Me offer the slightly better deal.

Classroom challenge

1 Gordon's Gourmet Store sells tubs of 'Posh Parsnips' at 5 tubs for £2·10.
Pierre's Perfect Food Store sells the same tubs at 3 tubs for £1·32.
Which store offers the better value?

2 Tasko sell a pack of 9 toilet rolls for £5·04.
Oldo sell the same toilet rolls in packs of 4 at a cost of £2·44.
Which shop offers the better deal?

3 Sweet potatoes are on sale in a farm shop at £20 for a 12·5 kilogram bag.
The same type of sweet potatoes are on sale in a supermarket at £3·80 for a 2·5 kilogram bag.
Which place offers the better value?

4 Craig is looking at bottles of 'Metal Brew' juice on the shelf of a shop.
A 1·5 litre bottle sells for £1·80.
A 250 ml bottle sells for £0·32.
Which is the better value?

5 Cola is on offer at Ladle and Sunsberry.

Ladle	2 litre bottles	offer	£1·32 each	or 3 for £3·60.
Sunsberry	250 ml cans	offer	18p each	or a pack of 24 cans for £4.

Which is the better value?

6 2 bottles of 'Hairshine' shampoo are on sale as shown.

Which is the better deal?

7 A teacher wants to buy 21 sweatshirts for students going on a school trip.
The sweatshirts normally cost £20 each.
The teacher sees these two offers.
From which shop should the sweatshirts be bought?

8 Carry's electrics are offering a 30% discount on a computer tablet that normally sells for £240.
CP World is selling the same tablet, for £190 + VAT at 20%.
Which shop offers the better value?

9 Nikki is buying tickets for a concert.
She sees two websites offering the tickets.
They advertise the costs as shown.
Nikki wants to buy a ticket. The ticket price is £80.
Which website offers the better value?

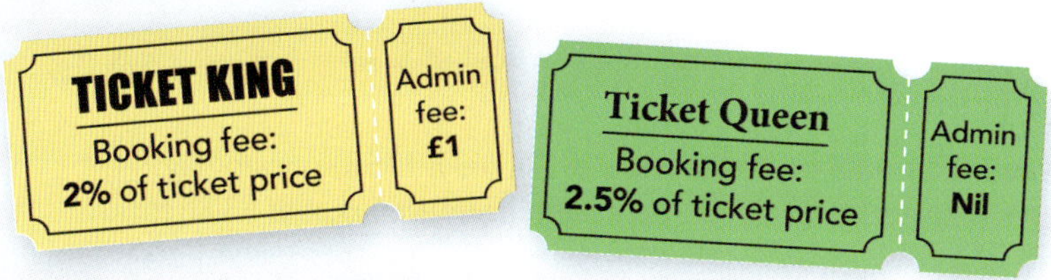

10 Clark visits a stationery shop and sees a number of offers.
For each offer in the table below, say whether Option 1 or Option 2 is the better value.
The first one has been done for you.

	Option 1	Option 2	Rate Option 1	Rate Option 2	Better option
1	3 batteries for £4·80	12 batteries for £14·76	£4·80 ÷ 3 £1·60	£14·76 ÷ 12 £1·23	Option 2
2	20 staplers for £330	5 staplers for £80			
3	5 calculators for £85	9 calculators for £155·70			
4	18 pens for £6·84	24 pens for £9·84			
5	11 notepads for £9·90	40 notepads for £44·80			
6	15 highlighters for £12·30	25 highlighters for £19·75			

11 Karen's Carpenters charge a rate of £36 per hour plus a call-out fee of £30.
Jake's Joiners charge a rate of £27 per hour and a call-out fee of £35.
Ben wants to floor his attic and asks the two companies for quotes.
Karen's Carpenters quoted Ben seven hours to do the job.
Jake's Joiners quoted Ben nine hours to do the same job.
Which company gave the lower quote, and by how much?

12 Jamie wants to install new satellite TV equipment in his house.
He checks out two companies.
SATtv charge £35·50 per hour, and have a call-out fee of £120.
Edge Media charge £42 per hour and a call-out fee of £90.

Both companies quote 3 hours to install the equipment.
Which company offers the better deal, and by how much?

13 Mike is a motor electrician.
He has a call-out charge of £50 and then charges £45 per hour when doing repairs.
Dawn is also a motor electrician.
She does not have a call-out charge, but charges £60 per hour when doing repairs.
- a If a repair took 2 hours, who would be cheaper?
- b If a repair took 3 hours, who would be cheaper?

14 Aspen is studying three different mobile phone providers' tariffs.

Company	Cost per month	Call minutes	Texts	Internet
DD	£15	200 free – then 12p per minute	300 free – then 5p per text	unlimited
P5	£27·50	400 free – then 8p per minute	100 free – then 3p per text	unlimited
Lemon	£42	500 free – then 8p per minute	Unlimited	Unlimited

Last month, Aspen's statement said she had used 420 minutes of calls and 250 texts.
- a How much would it have cost Aspen last month if she had chosen DD?
- b Would it have been cheaper for Aspen if she had chosen P5?
  Aspen decided to take out a contract with P5.
  The following month she used 500 minutes of calls and sent 350 texts.
- c Would it have been better if she had chosen Lemon?

15 Here are three bottles of brown sauce.
Which one offers the best deal?

200 ml
£1·30

300 ml
£1·92

500 ml
£3·15

Investigation

In small groups ...
Choose a company that provides energy, gas and electricity.
Try to compare their tariffs – charges – for energy.
Consider whether you only use one type of fuel or if you have dual fuel.
Does it make a difference if you pay by direct debit?
Does the company charge if you wish to change provider?
Can you save money if you have a 'Smart' meter?
Write a report, or make a presentation, to show others in the class.

Budgeting

Effective budgeting and financial vocabulary

I can budget effectively, making use of technology and other methods, to manage money and plan for future expenses. MNU 3-09b

What's coming up?

This Outcome and Experience will give you the opportunity to:

- budget effectively, using digital technology where appropriate, showing development of financial capability
- demonstrate knowledge of financial terms, for example, debit/credit, APR, p.a., direct debit/standing order and interest rate.

What you already know

You have already learned how to:

- ✔ carry out money calculations involving the four operations
- ✔ compare costs and determine affordability within a given budget
- ✔ demonstrate understanding of the benefits and risks of using bank cards and digital technologies.

In everyday life it is essential that you can 'manage your finances' effectively.

It is also important that you know what all the financial terms mean so that you can make informed decisions.

How does that work?

The table shows a list of commonly used words when dealing with money and finance.

You should aim to be familiar with most of them!

Word	Meaning
AER	Annual equivalent rate. This indicates how much interest is to be charged
APR	Annual percentage rate. This indicates how much interest is to be charged
ATM	Automated telling machine – sometimes called 'Hole in the wall'
Banknote	Pieces of 'paper' money, for example, the £20 note
Borrow	To get money from, e.g., a bank which must be paid back
Budget 'To budget'	An amount of money set aside for a purpose To check income and expenditure to determine a surplus or deficit

Word	Meaning
Cash	Actual money, coins or banknotes, paid. Not credit
Currency	The money used in a country. For example £ pound, € euro, $ dollar
Credit	Used to pay for goods over a period of time
Credit card	Card issued in order to purchase goods – a form of 'borrowing' – so that actual cash is not handed over
Debt	Money owed by one person to another
Debit card	Allows access to your bank account
Deficit	When more money is 'going out' than 'coming in'
Deposit	Money paid before receiving goods on credit or hire purchase Money paid into a bank account
Direct debit	Can be set up on your bank account to pay an amount to a company or a person The amount can be varied
Exchange rate	The rate at which one currency can be exchanged for another For example, £1 = €1·11, or £1 = 16·8 South African rand
Fee	Payment made to a professional, e.g. a solicitor Payment made for a service, e.g. a booking fee when booking online
Interest	Money paid for borrowing or investing money
Interest rate	The rate (a percentage) at which interest is charged or given
Loan	Money borrowed which must be repaid
Loss	When you sell an item for less than you paid for it
p.a.	'Per annum'– each year
Profit	When you sell an item for more than you paid for it
Refund	Pay back of money, e.g. if goods are returned
Standing order	Set up a fixed amount to be transferred from your account to a company or person
Surplus	When more money 'coming in' than 'going out'
Withdraw	To take money out, e.g. at an ATM or a bank

There are a several other money or financial words. You may wish to look some up and add them to this list.

When saving up to buy goods you should look at your budget to see how long it may take before you have enough – or if there are places where you can spend less money!

EXAMPLE

Marilyn earns £1250 per month.
She writes down a list of what she spends each month.

a Does Marilyn have a ***surplus*** or a ***deficit*** each month?

b How much is her surplus/deficit?

c Marilyn wants to save up £200 for a new tablet computer. How many months will it be before she has saved enough?

Item	Amount £
Rent for flat	£585
Council tax	£144
Electricity and gas	£142
Car insurance/tax	£45
Fuel for car	£80
Food/groceries	£140
Socialising	£50

SOLUTION

a Income £1250 Outgoings £1186 (add up all the amounts)
Income is more than outgoings – more money 'coming in' than 'going out'.
So, a ***surplus***

b surplus is £1250 – £1186 = £64

c Saves £64 per month £200 ÷ 64 = 3·125 therefore, she must save for 4 months
3 months would not be quite enough.

Classroom challenge

1 Copy and complete this table. The first line has been done for you.

	a	b	c	d
Income	£230	£145	£357·60	£62·43
Expenditure	£190	£163	£315·28	£62·43
Surplus	£40			
Deficit	–			

Mortgage	£349
Council Tax	£98
Gas	£55
Electricity	£62
Car loan payment	£181
Car insurance/tax	£32
Car fuel	£75
Food	£125
Entertainment	£53

2 Craig is hired as an admin assistant in a school.
The money he takes home each month is £1249.
He writes down his monthly bills.

a Does Craig have a surplus or deficit?

b How much is his surplus/deficit?

c How many months will it take Craig to save up for a £1000 holiday, if he puts all of his surplus towards the holiday?

3 Monica keeps track of her income and expenditure over four weeks.
This is her record.

Week	Income	Expenditure
1	£53	£35·50
2	£41·50	£38·25
3	£32·85	£29·15
4	£35·60	£15·60

a Calculate Monica's total income for the four weeks.

b Calculate Monica's total expenditure for the four weeks.

c Calculate Monica's surplus.

4 Elaine is a plumber. She earns £345 per week.
She makes a note of her outgoings for the week.

Rent and council tax	£135·50
Travel costs	£37·85
Food	£64·12
Leisure centre membership	£15·00
TV, phone, internet	£26·50
Gas and electricity	£31·25

How much does Elaine have left for saving or other spending?

5 Last year, the Murphy family made a note of their monthly budget.

Income	Expenditure	Amount
£2800	Mortgage	£695
	Council tax	£130
	Bus Pass	£38
	Entertainment/going out	£122
	Food	£365
	Electricity	£135
	Insurance – house/contents	£70
	Media package	£65

This year the Murphy family's income has increased by 1%.
The expenditure has increased by 5%.

a What was the Murphy family's expenditure last year?
b What was their monthly surplus last year?
c What is the family income this year?
d What is the expenditure this year?
e What difference has this made to the surplus?

6 The Borrelli family are saving up for a holiday.
They have worked out the costs for the holiday and these are shown in the table.

Flights	£920
Accommodation	£725
Travel insurance	£65
Car hire	£200
Spending money	£1000

a What is the total budget set aside for the holiday?
b They have 30 weeks to save up the money they need.
How much must they save each week?

Borrowing and credit

Sometimes, if you want to buy a big or an expensive item, for example a car or a new fitted kitchen, you may need to borrow money to pay for it. Another way of buying goods is to buy 'on credit', where you can spread the cost over a number of months or years.

How does that work?

EXAMPLE

Yvonne wants to buy a dining set which costs £899.
The shop offers a credit deal where Yvonne would pay a deposit, or initial payment, of £100 and then pay £24·50 per month for 36 months.
How much extra would Yvonne pay if she bought the dining set on credit?

SOLUTION

Deposit		£100
36 payments of	£24·50 = 36 × £24·50	£882
Total paid		£982
Extra paid	£982 – £899	£83

Although Yvonne is paying more over a period of time, she may consider this as the full price is a lot of money to find all at once, whereas monthly payments may be easier to budget for.

EXAMPLE

Jamie wants to borrow £5000 to buy a car.
He sees two banks who are offering a £5000 loan on the following terms.
TriBank offer: Repay £105 per month for 5 years
OctoBank: Repay £165 per month for 3 years
Which is the better offer?

SOLUTION

TriBank: 5 years = 60 months £105 × 60 = £6300
In this offer Jamie would pay less per month, but the cost of borrowing £5000 would be £1300.
OctoBank: 3 years = 36 months £165 × 36 = £5940
In this offer Jamie pays more per month, but cost of borrowing £5000 would be £940.
So OctoBank is the better offer in terms of total cost.
However, Jamie may decide to choose TriBank as, although the total cost is more, it is spread over 5 years and so the monthly repayments may be more manageable.

Classroom challenge

1 Johnny wants to borrow £2500 to buy a new guitar.
He sees two loan offers.
Offer 1: Borrow £2500 plus a credit charge of 8%, over 1 year.
Offer 2: Repay £109 per month over 2 years.
 a Which is the better offer?
 b Why might Johnny choose the other offer?

2 Marco Pierre is buying a new kitchen. He sees two offers from local finance companies.
 a Which offer is the cheaper?
 b Why might Marco Pierre choose the other offer?

Lend-A-Hand	LOAN 2
£3200 loan	£3000
Repay £77 per month over 4 years	Repay £181 per month over 18 months

3 Rory gets a quote for his car insurance.
He can pay either:

- a one-off payment of £612·40
- 12 monthly payments of £54·10.

Which method do you think Rory should choose?
Why?

4 Marion has a balance of £350 on her credit card.
When the statement comes in at the end of the month it states:
Minimum payment due,
either £10
or 3% of balance whichever is the greater
What is the minimum amount Marion must pay?

5 Investigate the differences between a debit card and a credit card.
Copy the table and fill in the boxes to show the differences

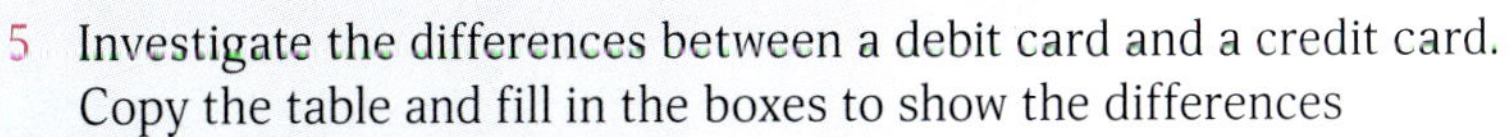

What to compare	Debit card	Credit card
What is it	A debit card is issued by a bank to a customer, who has an account with them It allows customers to purchase goods or services. The card is linked directly to the bank account	A credit card is issued by a bank (or other financial provider) to allow you to buy goods or services 'on credit' The bank pays on your behalf – and you then repay
When to pay		Later when the statement comes in, a payment is made. Either the full amount, or payments can be spread over a period of time
Bank Account	A bank account is essential for issuing a debit card	
Limit	Limit is amount of funds in linked account	
Interest		Interest will be charged if full payment is not made within agreed time period

6 Trust US Bank has different interest rates depending on how much you borrow.
The percentage rates they charge are shown in the table.

Loan amount	£1–£500	£501–£1000	£1001 – £2000	£2001–£5000	£5001 and over
APR: rate charged for 1 year	12%	9%	7%	5%	3%

a Dave wants to borrow £420, over 1 year, to buy a new bike.
 i What interest rate would Dave expect to pay?
 ii How much interest must he pay?
 iii What is the total he must repay?
 iv What would each monthly payment be?

b Colin wants to borrow £2400, over 1 year, to fit out his hairdressing salon.
 i How much interest will Colin be charged?
 ii What will be his monthly payments over the year?

c Sheena wants to borrow £850, over 1 year, to replace equipment in her fitness studio.
   i How much will Sheena have to pay back in total?
   ii Sheena budgets £80 per month to cover the repayments. Has Sheena set enough aside?

d Alanna wants to borrow £1000, over 1 year, to pay for a holiday. Her friend says she would be better to borrow £1001. Is her friend correct?

7 Mhairi wants to borrow £600 to buy a dress for the prom dance. The bank charges an interest rate of 9·2% over 1 year. Mhairi thinks she can afford to set aside £56 per month from her pay. Will this be enough to cover the monthly repayments?

Converting currencies

What's coming up?

This Outcome and Experience will give you the opportunity to:

- convert between different currencies.

What you already know

You have already learned how to:

- ✔ work with decimals including money
- ✔ calculate quantities which are in proportion.

In Britain, the currency we use is the GBP, the Great Britain pound.

Other countries will use their own currency.

For example, a number of European countries use the euro, €.

America uses the US dollar, \$. Denmark uses the kroner, South Africa uses the rand.

When you travel to other countries you will need to change GBP into the currency of the country.

Banks and bureaux de change will change pounds into a currency of your choice.

They will use an **exchange rate** to convert from pounds to the currency.

How does that work?

EXAMPLE

Sam is going to France on holiday.
She wants to exchange £250 for euros to spend whilst in France.
She notices that the exchange rate at the bank states that £1 = €1·14
How many euros will Sam get for her £250?

SOLUTION

She would get €285.

EXAMPLE

When Grant returns from South Africa he has ZAR250 left (ZAR is South African rand).
The bureau de change states the exchange rate of £1 = ZAR 16·7.
How many GBPs will Grant get for his rands?

SOLUTION

	ZAR	GBP £	
	16·7	1	
÷ 16·7			÷ 16·7
	1	0·05988	
× 250			× 250
	250	14·97	

Grant would get £14·97 for his rands.
In this type of question, the final answer would be rounded to the nearest penny.
When changing from foreign currency to GBP it is usually simpler to simply divide by the rate.
In the example above the working would be:

ZAR250   Exchange rate £1 = ZAR 16·7
250 ÷ 16·7 = 14·97   He would get £14·97.

Classroom challenge

In this exercise, use this currency conversion table.

Currency	Symbol	1 GBP (£1) buys ...
Euro	€	1·14
US dollar $	$	1·39
Australian dollar	A$	1·74
Danish kroner	Kr	8·45
South African rand	ZAR	16·87
Chinese yuan (Renminbi)	¥	8·92
Indian rupee	₹	88·96

1. Cheryl is going to Mijas, in Spain on holiday.
   She wants to change £500 into euros for spending money.
   Calculate how many euros Cheryl will get.
2. Keith is going to Las Vegas in the United States.
   He wants to exchange £1200 for US dollars.
   How many US dollars will he get?
3. A group of friends are going on safari in South Africa.
   The trip is going to cost £2100.
   They need to pay in South African rand.
   How many rand will they need to pay for the trip?
4. Albert is going to Odense, in Denmark.
   He intends to take the equivalent of £400 in Danish Kroner.
   How many kroners will Albert receive?
5. Craig and Charlie went to Australia for a holiday.
   They changed £1750 each into Australian dollars.
   How many dollars did they each get?
6. James saw a new mobile phone advertised for £350.
   How much would this be in:
   a Chinese yuan b Indian rupees c USA dollars d euros?
7. Mike saw a new digital camera priced at £740.
   How much would this be in:
   a Danish kroners
   b Australian dollars
   c USA dollars
   d South African rand?
8. Patricia saw a pottery kiln advertised, in Britain, for £830.
   She saw the same kiln advertised in the United States, for $1300.
   Which was the cheaper?

9 Harry returned from a trip to Bremen, in Germany, with 70 euros.
Using the exchange rate above, calculate how many GB pounds Harry got when he changed the euros back.

10 Whilst in South Africa, Patricia saw a necklace for ZAR3700.
Calculate how much this is in GBP.

11 Michael was on a trip to Beijing, in China.
Whilst there, he bought a hat costing 900 Chinese yuan.
Calculate how much this is in GBP.

12 The Smith family collected together some of the foreign currency they had left over after their holidays.
How many £ would they get when they exchanged these amounts of foreign currency?

a 80 euros
b US$ 160
c Australian $ 210
d 130 Danish kroner
e ZAR 1400
f 500 Indian rupees

13 Bart changed £500 to euros before going on holiday to Lisbon, Portugal.
Whilst there he spent €420.

a How many euros did Bart receive?
b How many euros did he bring back with him?
c How many £s would he get when he exchanges his remaining euros?

14 Miss Edmond is running a school trip to New York.
She wants to buy subway tickets, in advance.
A 7-day unlimited use ticket is $31.
There are 24 students and 3 teachers on the trip.

a What would be the total cost, in US dollars, for the 27 sets of 7-day tickets?
b How much is this in £s. Round your answer to the nearest £.

15 a A day ticket to the Taj Mahal costs 1000 rupees. How much is this in £?
b The Taj Mahal complex was completed in 1653 at an estimated cost of 32 million rupees.
What would this be in £?
c In 2015 it was estimated to be worth 52 billion rupees. How much is this in £?

16 When Gregor went to France on holiday he got an exchange rate of £1 = €1·14.
When he returned, he got an exchange rate of £1 = €1·18.
 a Gregor exchanged £250 to euros. How many euros did he get?
 b Whilst in France he spent €226. How many euros did he bring back?
 c How many GBPs would Gregor get for his remaining euros?

17 Cary went on holiday to Sydney, Australia.
She changed £2000 into Australian dollars at a rate of £1 = 1·72 Australian dollars.
Whilst in Sydney, she spent 3000 Australian dollars.
On her return, Cary exchanged her remaining dollars at a rate of £1 = 1·80 Australian dollars.
 a How many Australian dollars did Cary get for her £2000?
 b How many Australian dollars did she bring back?
 c How many GBP did she receive for her dollars?

18 The cost of a one-way ticket on the Tokyo–Kyoto 'Bullet Train' is 13 710 Japanese yen.
The exchange rate is £1 = 155·11 Japanese yen.
What is the cost of the ticket equivalent to in GBP?

MEASUREMENT SKILLS

TIME

Using simple time periods, I can work out how long a journey will take, the speed travelled at or distance covered, using my knowledge of the link between time, speed and distance. MNU 3-10a

What's coming up?

This Outcome and Experience will give you the opportunity to:

- apply knowledge of the relationship between speed, distance and time to find each of the three variables
- calculate time durations across hours and days.

What you already know

You have already learned how to:

- read and record time in both 12 hour and 24 hour notation and convert between the two
- know the relationships between commonly used units of time and carry out simple conversion calculations, for example, change $1\frac{3}{4}$ hours into minutes.
- use and interpret a range of electronic and paper-based timetables and calendars to plan events or activities and solve real-life problems
- calculate durations of activities and events including situations bridging across several hours and parts of hours using both 12 hour clock and 24 hour notation
- estimate the duration of a journey based on knowledge of the link between speed, distance and time
- choose the most appropriate timing device in practical situations and record times using relevant units, including hundredths of a second
- select the most appropriate unit of time for a given task and justify the choice.

Speed, distance and time

Working with hours and minutes

Unlike metric measures such as metres and centimetres, litres and millilitres, time is not 'decimalised'. This means it is not straightforward to use a calculator to help carry out time calculations.

EXAMPLE

We know that $2\frac{1}{2}$ hours can be written as 2·5 hours ($\frac{1}{2} = 0{\cdot}5$).
So we can use 2·5 on a calculator as it is in decimal form.
What about 2 hours and 24 minutes?

SOLUTION

2 hours and 24 minutes will be 2· 'something'.
We need to write 24 minutes as a decimal fraction of an hour.

24 minutes $= \frac{24}{60}$ hour $= 24 \div 60 = 0{\cdot}4$ hour

So 2 hours and 24 minutes = 2·4 hours.
So now a calculator could be used, as 2·4 hours is in decimal form.

EXAMPLE

What would 4·3 hours be in hours and minutes?

SOLUTION

4·3 hours = 4 hours and 'something' minutes
0·3 hour = 0·3 × 60 (there are 60 minutes in an hour)
= 18 minutes
4·3 hours = 4 hours 18 minutes

Classroom challenge

1 Convert these hours and minutes to hours. Where appropriate, round your answer to two decimal places.

a 1 hour 30 minutes
b 2 hours and 15 minutes
c 4 hours 45 minutes
d 3 hours 6 minutes
e 5 hours and 36 minutes
f 8 hours 12 minutes
g 0 hours 48 minutes
h 6 hours 20 minutes
i 7 hours 40 minutes
j 3 hours 3 minutes
k 12 hours 18 minutes
l 2 hours 25 minutes

2 Write these hours as hours and minutes.
Where appropriate, round your answer to the nearest minute.

a 1·5 hours
b 3·25 hours
c 6·75 hours
d 3·2 hours
e 6·7hours
f 5·45 hours
g 9·3 hours
h 12·78 hours
i 1·15 hours
j 6·64 hours
k 3·82 hours
l 4·04 hours

Calculating time taken for a journey

How does that work?

EXAMPLE

A car travels 80 kilometres at an average speed of 20 km/hr. How long will it take?
20 km/hr means that the car travels 20 kilometres in one hour. The table shows the distances covered for different times at this speed.

Speed	Time	Distance
20 km/hr	1 hour	20 km
20 km/hr	2 hours	40 km
20 km/hr	3 hours	60 km
20 km/hr	4 hours	80 km
20 km/hr	5 hours	100 km

From the table, to travel a ***distance*** of 80 km at a ***speed*** of 20 km/hr will take a ***time*** of 4 hours.

We can use the formula time taken = distance ÷ average speed

This is usually written as T = D ÷ S or $T = \frac{D}{S}$

EXAMPLE

Mike cycles 45 km at an average speed of 15 km per hour.
How long will Mike's cycle ride take?

SOLUTION

$$T = \frac{D}{S}$$

$$= \frac{45}{15} \quad \text{or} \quad 45 \div 15$$

$$= 3 \text{ hours}$$

DON'T FORGET

Check that units are alike. Here the distance is in **kilometres** and the speed is in **kilometres** per hour. So the answer will be in hours.

EXAMPLE

Debbie is steering her narrow boat along a canal that is 24 kilometres long.
Her average speed is 3·6 km/hr.
How long will Debbie take to complete the canal trip?
Give your answer in hours and minutes.

SOLUTION

$$T = \frac{D}{S}$$
$$= \frac{24}{3\cdot6} \quad \text{or} \quad 24 \div 3\cdot6$$
$$= 6\cdot6667 \text{ hours} \quad \text{or} \quad 6\tfrac{2}{3} \text{ hours}$$
$$= 6 \text{ hours } 40 \text{ minutes}$$

Classroom challenge

1 The table shows the distance travelled and the average speed of various vehicles.
For each one, calculate the time taken.
Where appropriate, give your answer in hours and minutes, to the nearest minute.

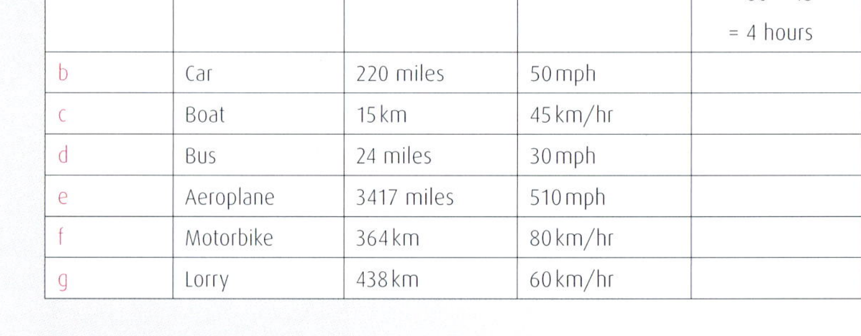

Question	Vehicle	Distance	Average speed	Time taken
a	Bicycle	60 km	15 km/hr	$T = D \div S$ $= 60 \div 15$ $= 4$ hours
b	Car	220 miles	50 mph	
c	Boat	15 km	45 km/hr	
d	Bus	24 miles	30 mph	
e	Aeroplane	3417 miles	510 mph	
f	Motorbike	364 km	80 km/hr	
g	Lorry	438 km	60 km/hr	

2 Grace rides her horse at an average speed of 8 kilometres per hour.
How long will she take to travel a distance of 18 kilometres?

3 Karen can skate her skateboard at an average speed of 6 mph.
How long will it take Karen to skate 9·6 miles?

DON'T FORGET

For these questions look at the units carefully.

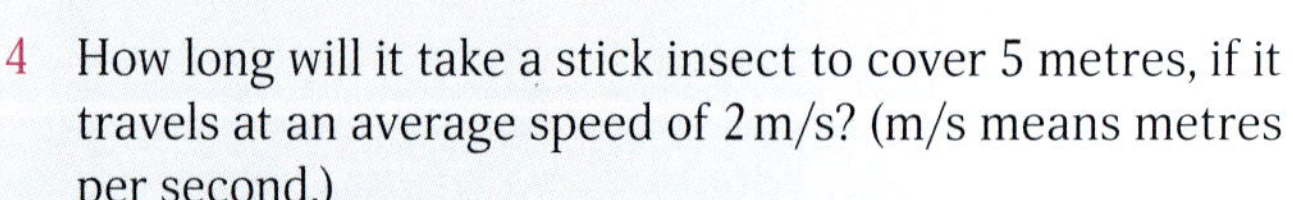

4 How long will it take a stick insect to cover 5 metres, if it travels at an average speed of 2 m/s? (m/s means metres per second.)

5 Steven is going from his art class to his mathematics class, a distance of 145 metres.
If he can walk along the corridors at 1·6 m/s, how long will it take Steven to get to his class?

6 A discus thrower threw the discus a distance of 142·6 metres.
Whilst in the air, the discus travelled at an average speed of 12·4 m/s.
How long was the discus in the air?

7 A 'Segway® Personal Transporter (PT)' travels 4·5 km at a speed of 9 km/hr.
A cyclist travels the same distance at 12 km/hr.
How much less time did the cyclist take?

Calculating the average speed on a journey

How does that work?

EXAMPLE

John travels 80 km in 4 hours.
What was his average speed?

SOLUTION

Average speed is given in units such as kilometres **per** hour, miles **per** hour, metres **per** second, feet **per** second

Per means 'for each; that is we divide.
John travels 80 kilometres in 4 hours.
To find how far he travels in 1 hour we must divide.

80 ÷ 4 = 20 kilometres per hour
= 20 km/hr

We can use a formula like the one for time:

speed = distance ÷ time or
S = D ÷ T or

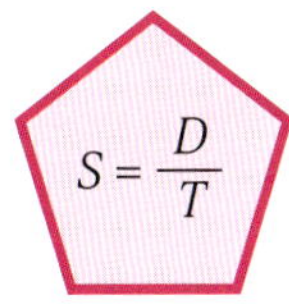

EXAMPLE

Mark travels 120 km in a time of 3 hours and 45 minutes.
What is Mark's average speed?

$$S = \frac{D}{T} \quad \text{or } D \div T$$
$$= \frac{120}{3{\cdot}75} \quad \text{or } 120 \div 3{\cdot}75$$
$$= 32$$

Note: the 45 minutes must be changed to a decimal fraction of an hour. 45 mins = 0·75 hours

Mark's average speed is 32 km/hr.

DON'T FORGET

Note that distance was in	kilometres
Time was in	hours
So, speed is in	kilometres per hour

Classroom challenge

1 Calculate the average speed of a car which travels 240 kilometres in 4 hours.

2 Calculate the average speed of a cyclist who travels 54 kilometres in 4 ½ hours.

3 Cindy jogged a distance of 15 kilometres in a time of 2 hours 30 minutes.
What was Cindy's average speed?

4 Calculate the average speed for each of these journeys.

	a	b	c	d	e
Distance	315 km	84 metres	467·5 miles	34 850 km	8 metres
Time	5 hours	14 seconds	8 hr 30 min	10 hr 15 min	2·4 minutes

5 Greg's journey to the airport took him 2 hours 30 minutes.
The distance to the airport is 120 miles.
Calculate Greg's average speed in miles per hour.

6 Meghan is delivering parcels for her courier firm.
She records the distances and the time taken between the towns on her route.
These are shown in the diagram.

Calculate Meghan's average speed for each part of the journey.

7 Shelley set off from home at 2pm to drive 120 km to visit her grandfather.
She arrives at 3.20pm.
- a How long did the journey take Shelley?
- b Calculate Shelley's average speed for the journey.
- c It took Shelley 1 hour 40 minutes to drive home.
What was Shelley's average speed for the return journey?

Calculating the distance travelled on a journey

How does that work?

EXAMPLE

Marina is driving along the M6 motorway at an average speed of 60 miles per hour.
How far will Marina travel in $4\frac{1}{2}$ hours?

Note:
- speed is in **miles** per **hour**
- time is in **hours**
- distance will be in **miles**.

SOLUTION

A speed of 60 miles per hour
means 60 miles travelled for **each** hour
in $4\frac{1}{2}$ hours $= 60 \times \mathbf{4\frac{1}{2}}$
$= 270$ miles

The formula here is distance = speed × time
or $D = S \times T$ $(D = ST)$

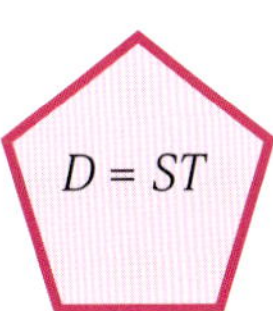

EXAMPLE

Abbey is driving at 80 km/h.
How far will she travel in 3 hours and 42 minutes?

SOLUTION

Step 1 Change the hours and minutes to a decimal 3 hours 42 minutes
As a decimal part of an hour 42 ÷ 60 = 0·7 Therefore 3·7 hours

Step 2 Solve as above
A speed of 80 kilometres per hour
means 80 kilometres for each hour
in 3·7 hours $= 80 \times \mathbf{3{\cdot}7}$
$= 296$ kilometres

The formula here is distance = speed × time
or $D = S \times T$ $(D = ST)$

Note:
- speed is in **kilometres** per **hour**
- time is in **hours**
- **distance** will be in **kilometres**.

Classroom challenge

1 A car travelled at 60 miles per hour.
How far would the car travel in:
a 1 hour b 3 hours c $5\frac{1}{2}$ hours?

2 Elaine ran at a speed of 6 metres per second.
How far would she run in:
a 1 second b 5 seconds c 10 seconds?

3 In this question, work out how far each vehicle would travel in the given time.

Question	Vehicle	Average Speed	Time	Distance
a	Bicycle	15 km/hr	3 hours	$D = S \times T$ $= 15 \times 3$ $= 45$ km
b	Car	50 mph	$4\frac{1}{2}$ hours	
c	Boat	45 km/hr	$1\frac{1}{4}$ hours	
d	Bus	30 mph	45 minutes	
e	Aeroplane	510 mph	6 hours 12 minutes	
f	Motorbike	80 km/hr	2 hours 48 minutes	
g	Lorry	60 km/hr	5 hours 20 minutes	

4 A centipede moves at 4 centimetres per second.
How far would it move in 40 seconds?

5 How far would an aeroplane fly, at 540 miles per hour, if it was in flight for 5 hours?

6 A fire engine was on an emergency call.
It was travelling at 65 km/hr.
The fire engine took 24 minutes to reach the scene.
a What fraction of an hour is 24 minutes?
b How far did the fire engine travel to the emergency?

7 Corinne delivers car parts to garages in various towns around the country.
Her route one day is shown below.

Her average speed and time taken for each section of the journey are shown.
a Calculate the distance between each town
b What was the total distance covered by Corinne?
c Copy and complete this distance table for the towns

Brechin			
	Montrose		
105 km		Stonehaven	
			Inverary

STRETCH YOURSELF

8 A cheetah, the fastest land animal, can reach a speed of around 115 km/hr in 'full flight'.
In 'full flight', how far would a cheetah run in 30 seconds?

Mixed examples of speed, distance and time

We now have three formulae for calculating each of the three measures; speed, distance and time.

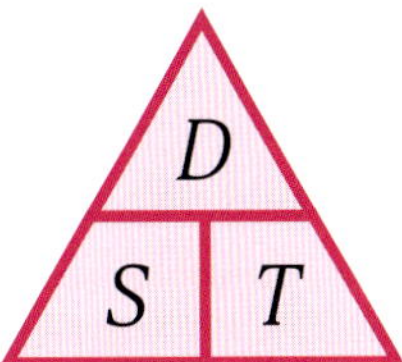

We can use this triangle to help us remember the three formulae below.

If you want to find the **time** taken; put your thumb over the '*T*'.

You are left with $\frac{D}{S}$ that is, $T = \frac{D}{S}$

If you want to find the **average speed**; put your thumb over '*S*'.

You are left with $\frac{D}{T}$ that is, $S = \frac{D}{T}$

If you want to find the **distance** travelled; put your thumb over '*D*'.

You are left with *S T* that is, $D = S \times T$

When answering questions on speed, distance and time, you should always take 3 steps:

Step 1 Write down the correct formula.

Step 2 Substitute the numbers – being very careful to check units are consistent.

Step 3 Write your answer – making sure to include the unit.

How does that work?

A bus travels between Musselburgh and North Berwick, which are 33 km apart.
The bus travels at an average speed of 44 km/hr.
How many minutes did the bus journey take?

SOLUTION

We are trying to find time taken.

Step 1 Formula $T = \frac{D}{S}$ or $D \div S$

Step 2 Substitute numbers $T = \frac{33}{44}$ or $33 \div 44$

$T = 0{\cdot}75$

Step 3 Write answer Time taken $= 0{\cdot}75$ hours

$= 45$ minutes (asked for answer in minutes)

DON'T FORGET

To change a decimal fraction of an hour into minutes, simply multiply by 60 (As there are 60 minutes in one hour). In this case $0{\cdot}75 \times 60 = 45$

Classroom challenge

1 Copy this table and fill in the missing parts.
Be careful to check the units.

Question	Vehicle	Average speed	Time	Distance
a	Bicycle	15 km/hr		90 km
b	Car		4 ½ hours	540 km
c	Boat	45 km/hr	2 ¼ hours	
d	Bus	50 mph	45 minutes	
e	Aeroplane	510 mph		3672 miles
f	Motorbike		2 hours 48 minutes	218·4 km
g	Lorry	60 km/hr	3 hours 40 minutes	

2 Eloise is driving along an autoroute in France.
She sees this sign.
Eloise reckons she can drive at an average speed of 90 km/hr.
How long will Eloise take to reach Dijon?

3 The distance from Bridgeton to Castleton is 30 kilometres.
The distance from Castleton to Drummond is 52 kilometres.

a Habib leaves Bridgeton, at 9:00, to deliver groceries in Castleton.
Habib drives at an average speed of 50 km/hr.
When does Habib arrive in Castleton?

b Habib leaves Castleton at 9:52.
She needs to be at Drummond for 10:40.
Calculate the average speed at which Habib would need to drive to arrive on time.

4 A helicopter is flying at a speed of 520 mph for 3 ½ hours.
How far will the helicopter travel?

5 Conrad is running in a marathon, a distance of 26·2 miles.
He wants to complete the marathon in 4 hours.
What average speed should he maintain to achieve this?

6 Patricia, Mike, Alan and Cheryl went on holiday to South Africa.
Mike kept a note of some of the activities they did whilst on holiday.

a They flew from Edinburgh to Doha, in Qatar.
The flight took 7 hours and 45 minutes.

The distance from Edinburgh to Doha is about 5580 kilometres.
What was the average speed of the plane?

b They then flew from Doha to Johannesburg in South Africa.
The distance from Doha to Johannesburg is about 6239 kilometres.
The plane flew at an average speed of 734 kilometres per hour.
How long did the flight take?
Give your answer in hours and minutes.

c Whilst in South Africa, the two couples flew from Johannesburg to Cape Town.
The flight lasted 2 hours 9 minutes.
The average speed of the plane was 580 kilometres per hour.
What is the distance from Johannesburg to Cape Town?

d Whilst in Cape Town, they took a cable car to the top of Table Mountain.
The ticket had the following information.

i How long would it take the cable car to get to the top of Table Mountain, at its maximum speed?
Give your answer in minutes.

ii On one trip, the cable car took 5 minutes to reach the top.
What was its average speed on this trip?
Give your answer in metres per second.

e Another part of the holiday involved driving from Johannesburg to the Nambiti Game Reserve.
The distance from Johannesburg to Nambiti Game Reserve is 374 kilometres.
Patricia, Mike, Alan and Cheryl needed to be there at 2pm to check in.
They reckoned they could travel at an average speed of 88 km/hr.
The also wanted to stop for 45 minutes for a coffee break.
At what time should they leave Johannesburg to arrive in time?

f Whilst on a game drive, they saw a lion chasing a herd of Wildebeeste.
Their guide reckoned the lion covered 300 metres in 15 seconds.
What was the average speed of the lion:

i in metres per second

ii in kilometres per hour?

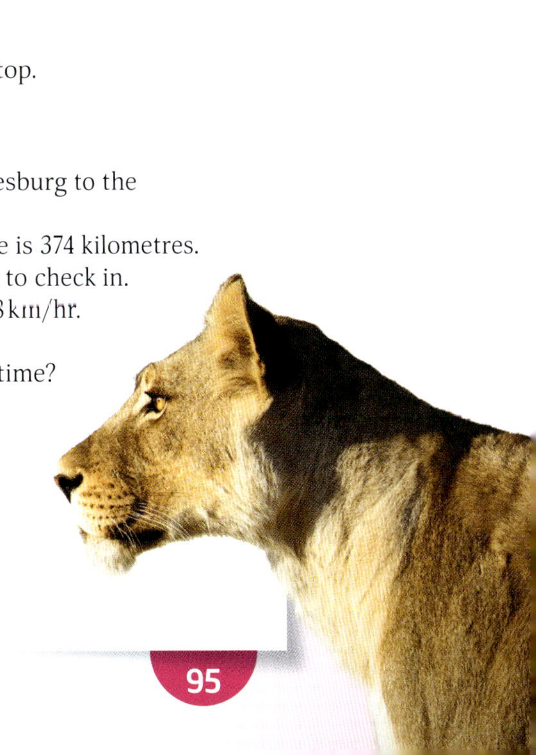

7 Fred and Barney are driving along the M8 motorway.
They see this sign.
Fred says, 'We will not reach the junction in that time, unless we break the speed limit'.
Is Fred right?
Show all your working.

M8 Motorway
Junction 6 32 miles
Time to Junction 28 minutes

DON'T FORGET

Motorway speed limit is normally 70 mph.

8 On the 25th August, 1875, Matthew Webb made the first successful swim across the English Channel.
The width of the Channel, where he swam was 34 kilometres.
Matthew swam the distance in 21 hours 45 minutes.
What was his average speed?

9 The distance from the Earth to the Moon is about 384 400 km.
The Space Shuttle can travel at an average speed of 28 000 km/hr.
How long will it take the Space Shuttle to reach the Moon?
Give your answer to the nearest hour.

10 On October 4th, 1983, Thrust 2, driven by Richard Noble, set the land speed record of 1020 km/hr. The record is set over a distance of 1 km.
If this was the average speed, how long would it take to cover 1 km?
Give your answer in seconds.

Time intervals

When planning journeys, it is useful to be able to calculate time intervals, or time differences.

How does that work?

EXAMPLE

A plane leaves Heathrow Airport at 22:30 (GMT) and arrives at JFK airport, in the USA, at 04:15 (GMT) the following day. How long did the flight take?

SOLUTION

Make a time line.

10:30	→	11:00	→	4:00	→	4:15
	30 min		5 hours		15 min	

Total time 5 hours 45 minutes.

EXAMPLE

Frank is baking a cake in the oven.
He knows it will take 1 hour 30 minutes.
After he takes it out, he needs to let it sit for 50 minutes to cool.
If he wants to serve it at 18:15, when should he put it in the oven?

SOLUTION

Total time taken 1 h 30 min + 50 min = 1 hr 80 min
= 2 hr 20 min

Count back from 18:15
6:15 – 2hours → 4:15 – 20 minutes → 3:55

Frank should put it in oven at 15:55.

Classroom challenge

1 Calculate the length of time between:

a	7:15am	and	9:23am
b	2:45pm		8:15pm
c	8:40am		3:24pm
d	11:52am		12:03pm
e	6:30am		7:05pm

2 Calculate the length of time between:

a	07:25	and	09:15
b	11:35		15:42
c	17:45		23:18
d	12:00		20:20
e	00:01		23:59

3 A cruise ship leaves Newhaven harbour at 11:40 and arrives at Hull harbour at 08:20 the next day.
How long did the cruise ship take?

4 Calculate the time taken for each of these flights.

	Depart	Arrive
a	22:30	06:45
b	23:17	07:23
c	19:45	05:12
d	20:25	08:16
e	18:24	11:06

5 Ryan works part-time and needs to 'clock in' and 'clock out' so that the Office can work out the hours he works.
Copy and complete this time sheet for Ryan.

Ryan Employee number 12876 Week number 16			
	Time in	Time out	Hours worked
Monday	8:30 am	2:00 pm	
Tuesday	9:40 am	2:30 pm	
Wednesday	8:15 am	3:15 pm	
Thursday	10:45 am	4:15 pm	
Friday	9:00 am	1:15 pm	
		Total hours worked	

6 A search and rescue boat was 'scrambled' at 22:45 in answer to a distress call from a fishing boat, who had an injured crewman on board.
The rescue boat reached the fishing vessel at 00:25.
They took the injured crewman to the nearest hospital where he arrived at 03:20.

a How long did it take the rescue boat to reach the fishing boat?

b How long after the distress call was the crewman admitted to hospital?

7 A departure board at an airport was showing incoming flight delays.
Copy the table and complete the column to show the expected arrival time of each flight.

Arrivals

Flight no.	From	Due	Delay	Expected arrival time
AK 103	Paris	23:40	1 hr 20 min	
EQ 241	Hamburg	23:52	2 hr 15 min	
BW 101	Malaga	20:17	3 hr 40 min	
XT 314	Lisbon	21:52	5 hr 06 min	
TY 278	Amsterdam	22:48	3 hr 43 min	

MEASUREMENT

Units, areas and volumes

I can solve practical problems by applying my knowledge of measure, choosing the appropriate units and degree of accuracy for the task, and using a formula to calculate area or volume when required. MNU 3-11a

What's coming up?

This Outcome and Experience will give you the opportunity to:

- choose appropriate units for length, area and volume when solving practical problems
- convert between standard units to three decimal places and apply this when solving calculations of length, capacity, volume and area
- calculate the area of a 2D shape where the units are inconsistent.

What you already know

You have already learned how to:

- ✔ use the comparative size of familiar objects to make reasonable estimations of length, mass, area and capacity
- ✔ estimate to the nearest appropriate unit, then measure accurately: length, height and distance in millimetres (mm), centimetres (cm), metres (m) and kilometres (km); mass in grams (g) and kilograms (kg); and capacity in millilitres (ml) and litres (l)
- ✔ calculate the perimeter of simple straight-sided 2D shapes in millimetres (mm), centimetres (cm) and metres (m)
- ✔ calculate the area of squares, rectangles and right-angled triangles in square millimetres (mm^2), square centimetres (cm^2) and square metres (m^2)
- ✔ calculate the volume of cubes and cuboids in cubic centimetres (cm^3) and cubic metres (m^3)
- ✔ convert between common units of measurement using decimal notation, for example, 550 cm = 5·5 m; 3·009 kg = 3009 g
- ✔ choose the most appropriate measuring device for a given task and carry out the required calculation, recording results in the correct unit
- ✔ read a variety of scales accurately
- ✔ draw squares and rectangles accurately with a given perimeter or area.

Using appropriate units and converting between units of measure

When measuring, or calculating, lengths, areas or volumes, it is important that appropriate units are used. This makes the measurements 'more sensible' and more easily used.

How does that work?

For length:
If you were asked to measure the length of ...

a stamp you would probably use millimetres (mm)
an envelope you would probably use centimetres (cm)
a Post Office counter you would probably use metres (m)
the route taken by a Post Office van in one day you would probably use kilometres (km)

For area:
If you were asked to measure the area of ...

a stamp you would probably use square millimetres mm^2
an envelope you would probably use square centimetres cm^2
a Post Office floor you would probably use square metres m^2
a town covered by postal deliveries you would probably use hectares or square kilometres km^2

For volume or capacity:
If you were asked to measure the volume of ...

a dose of medicine you would probably use millilitres ml
a sports water bottle you would probably use centilitres cl
a car engine you would probably use cubic centimetres cm^3, cc or litres l
a swimming pool you would probably use cubic metres m^3

The main, metric, units used in measure are:

Length	Area	Volume
Millimetre (mm)	Square millimetre (mm^2)	Cubic millimetre (mm^3)
Centimetre (cm)	Square centimetre (cm^2)	Cubic centimetre (cm^3)
Metre (m)	Square metre (m^2)	Cubic metre (m^3)
	Hectare (ha) (=10000 m^2)	
Kilometre (km)	Square kilometre (km^2)	
For volume, measures of capacity include		
		Millilitre (ml) = 1 cm^3
		Centilitre (cl) = 10 ml or 10 cm^3
		Litre (l) = 1000 ml or 1000 cm^3

It is a useful skill to be able to change from one unit to another.

How does that work?

EXAMPLE

Change 47 millimetres to centimetres.
Look at this ruler.

You can see that 47 mm = 4·7 cm.
Note that there are 10 mm in 1 cm.

In a similar way 2·4 cm = 24 mm.

In length: 10 mm = 1 cm
100 cm = 1 m
1000 m = 1 km

Make a note of this diagram as a useful way of changing from one unit to another.

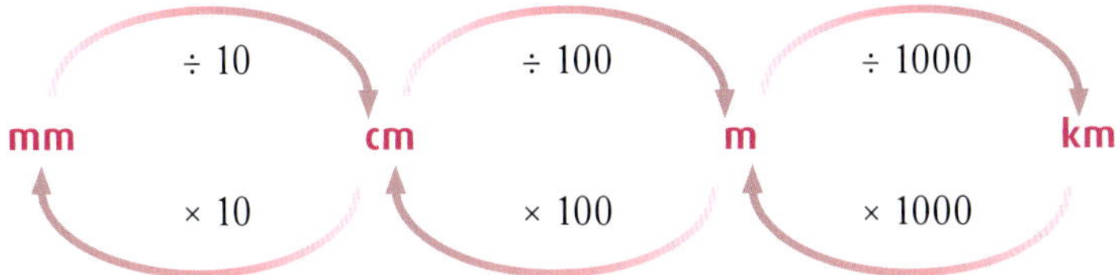

EXAMPLE

Change 3512 metres to kilometres	3512	÷	1000	=	3·512 km
Change 127 centimetres to metres	127	÷	100	=	1·27 m
Change 2 kilometres to metres	2	×	1000	=	2000 m
Change 2·5 centimetres to millimetres	2·5	×	10	=	25 mm
Change 3 metres to millimetres	3	×	100	=	300 cm
	300	×	10	=	3000 mm

For capacity, the conversion table looks like this:

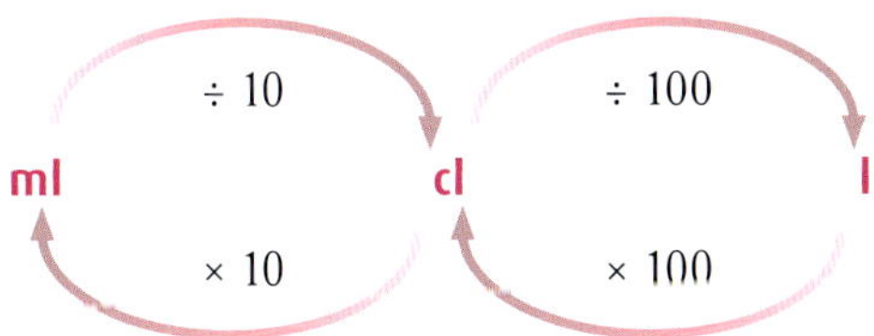

Converting from one unit to another. Use diagrams above or this table.			
Length	Area	Volume	Capacity
10 mm = 1 cm	100 mm² = 1 cm²	1000 mm³ = 1 cm³	10 ml = 1 cl
100 cm = 1 m	10000 cm² = 1 m²	1000000 cm³ = 1 m³	1000 ml = 1 l
1000 m = 1 km	10000 m² = 1 ha		1000 cm³ = 1 l
			1000 l = 1 m³

To change from small unit to large unit you must divide (think fewer larger units equal more smaller units).
To change from larger unit to smaller unit you must multiply (think more smaller units equal fewer larger units).

More conversions!

EXAMPLE

Convert $64\,000\,\text{cm}^2$ to m^2
$3200\,\text{cm}^3$ to litres
$0{\cdot}64\,\text{cm}^3$ to mm^3

SOLUTION

$64\,000\,\text{cm}^2$ to m^2	from the table above $64\,000 \div 10\,000$	$10\,000\,\text{cm}^2 = 1\,\text{m}^2$ $= 6{\cdot}4\,\text{m}^2$	**small unit to large unit:** $\div$
$3200\,\text{cm}^3$ to litres	from the table above $3200 \div 1000$	$1000\,\text{cm}^3 = 1\,\text{l}$ $= 3{\cdot}2\,\text{l}$	**small unit to large unit:** $\div$
$0{\cdot}64\,\text{cm}^3$ to mm^3	from the table above $0{\cdot}64 \times 1000$	$1\,\text{cm}^3 = 1000\,\text{mm}^3$ $= 640\,\text{mm}^3$	**large unit to small unit:** $\times$

Classroom challenge

1 Copy and complete these sentences, using an appropriate unit of measure.
   a I would measure the length of a flea in ....
   b I would measure the height of a door in ....
   c I would measure the width of my index finger in ....
   d I would measure the area of a stamp in .... ,
   e I would measure the area of a £10 note in ....
   f I would measure the area of a page of a newspaper in ....
   g I would measure the distance cycled in 3 hours in ....
   h I would measure the distance between Aberdeen and Stonehaven in ....
   i I would measure a dose of medicine in ....
   j I would measure the capacity of a small water bottle in ....
   k I would measure the capacity of a swimming pool in ....
   l I would measure the volume of a small jewellery box in ....
   m I would measure the area of my classroom floor in ....
   n I would measure the volume of my classroom in ....
   o I would measure the volume of my breakfast cereal box in ....
   p I would measure the area of my desk in ....
   q I would measure the length of my pencil in ....
   r I would measure the engine size of the school minibus in ....
   s I would measure the size of Murrayfield rugby pitch in ....
   t I would measure the area of a large farm in ....

2 Write each of the following in centimetres.

a	5 cm 4 mm	d	243 mm	g	6 m 4 cm
b	23 cm 7 mm	e	2 m	h	15·3 m
c	72 mm	f	4 m 17 cm	i	2018 mm

3 Write each of the following in millimetres.

a	4 cm	c	3 cm 5 mm	e	183 cm
b	6·2 cm	d	3 m	f	1 km

4 Write each of the following in metres.

a	5 m 19 cm	c	5 km	e	4 m 7 mm
b	6 m 8 cm	d	8 m 250 mm	f	3 km 72 m

5 Write each of the following in kilometres.

a	1 765 m	c	25 m	e	6 km 62 m
b	432 m	d	3 km 482 m	f	150 000 cm

6 Copy each of these sentences and fill in the blanks.

a	4000 millilitres is the same as ..... litres	c	35 000 ml is the same as ..... litres
b	..... litres is the same as 750 ml	d	400 cl is the same as ..... litres

7 Write each of the following in cm^2.

a	3 m^2	c	75 m^2	e	540 mm^2
b	5 m^2	d	3·141 m^2	f	1 672 mm^2

8 Write each of the following in mm^2.

a	4 cm^2	c	5·3 cm^2	e	0·345 cm^2
b	9 cm^2	d	42 cm^2	f	1·037 cm^2

9 Write each of the following in m^3.

a	6 000 000 cm^3	c	17 500 000 cm^3	e	12 000 cm^3
b	7 525 000 cm^3	d	245 000 cm^3	f	4350 cm^3

10 Write each of the following in mm^3.

a	2 cm^3	c	4·5 cm^3	e	0·37 cm^3
b	9 cm^3	d	0·4 cm^3	f	0·01 cm^3

11 Write each of the following in litres.

a	7000 cm^3	c	400 cm^3	e	2·175 m^3
b	16 500 cm^3	d	4 m^3	f	1·642 m^3

12 A tennis court is 28 m long and 8 metres wide.

a Calculate its area in m^2.

b The ball boys at a tennis match rest their knees on a pad which has an area of 0·54 m^2. How many cm^2 is this?

13 A net on a tennis court is set at a height of 1·07 m. The supporting poles are set 0·914 m outside the court line.

a Write the height of the net in centimetres.

b Write the distance of the poles from the court line in centimetres.

14 Manor Farm covers an area of 43 hectares. How many square metres is this?

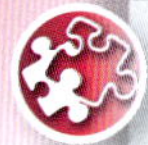

15 The volume of an Olympic size swimming pool is 2500 m³. Write this in litres.

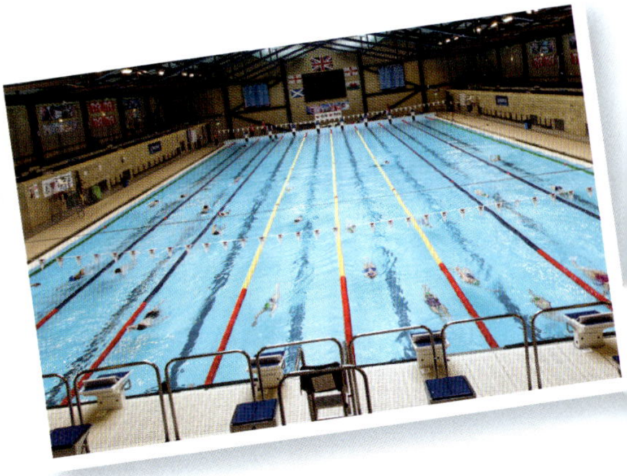

16 A doctor tells Judy 'take a 5 ml spoonful of medicine three times per day'.
The medicine bottle Judy has contains 0·2 l of medicine.

a How much medicine does Judy take each day?

b Will Judy have enough medicine for two weeks?

Area of 2D shapes

Name of shape	Diagram	Formula for area
Rectangle	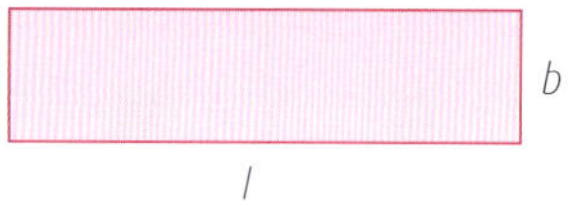	Area = length × breadth $A = lb$
Square	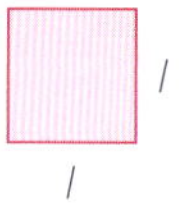	Area = length × length $A = l^2$
Triangle	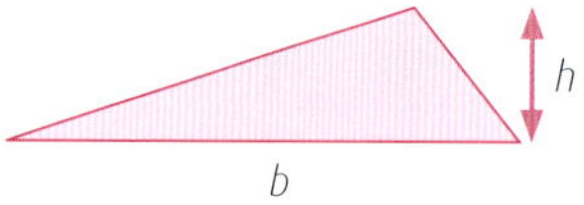	Area = $\frac{1}{2}$ base × height $A = \frac{1}{2}bh$

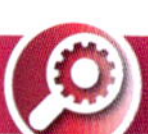

How does that work?

Always use '***Formula*, *substitute*, *calculate***'

Calculate the area of these shapes.

1

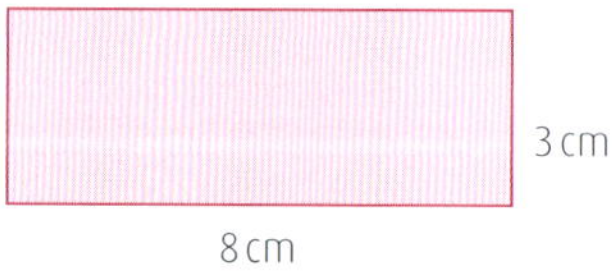

A	=	lb	formula
	=	8×3	substitute
	=	24 cm^2	calculate

2

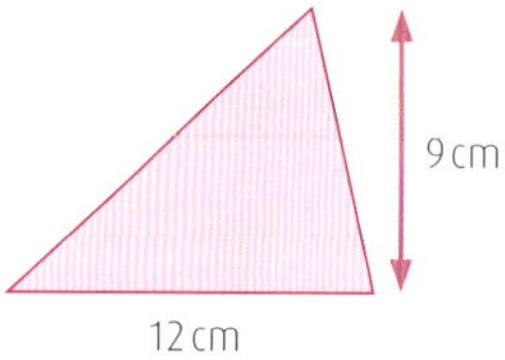

A	=	$\frac{1}{2}bh$	formula
	=	$\frac{1}{2} \times 12 \times 9$	substitute
	=	54 cm^2	calculate

3 The diagram shows a farmer's field.
The field is 1·5 km long and 600 m broad.
What is its area, in hectares?

A	=	lb	formula
	=	1500×600	substitute
	=	$900\,000\ \text{m}^2$	calculate
In hectares	=	$900\,000 \div 10\,000$	
	=	90 hectares	

Careful! Make sure units are the same.

Classroom challenge

1 Calculate the area of these rectangles and squares.

a

7 cm
3 cm

b

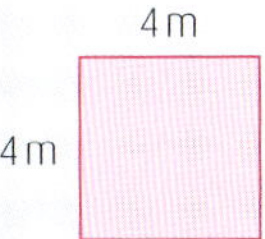

c

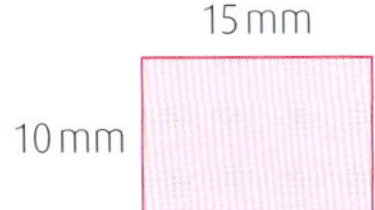

2 Calculate the area of these triangles.

a

19 m
18 m

b

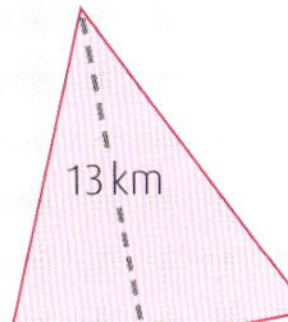

c

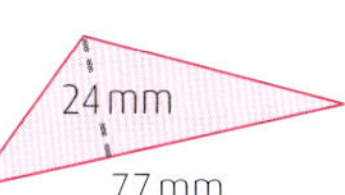

d

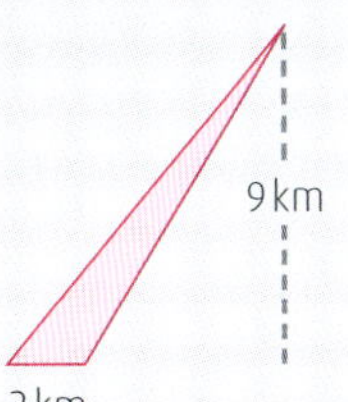

e

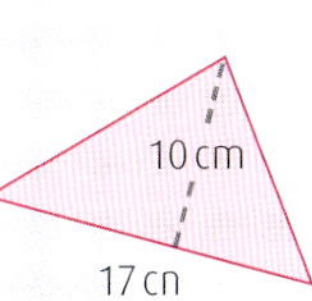

f

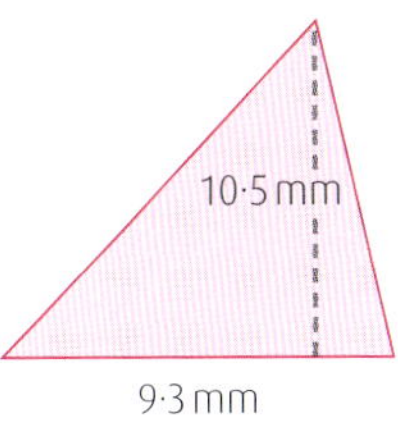

g

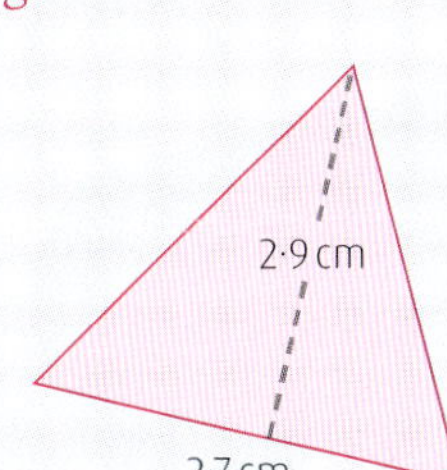

h

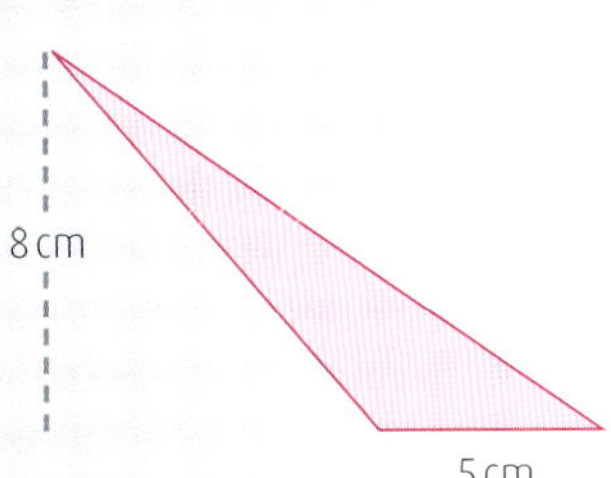

i

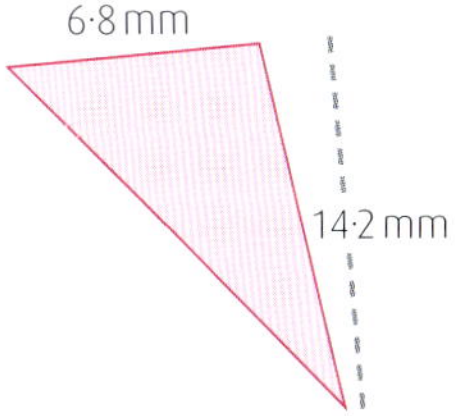

3 Calculate the area of these shapes.

a This table top is in the shape of a rectangle. Calculate the area of the table top in cm².

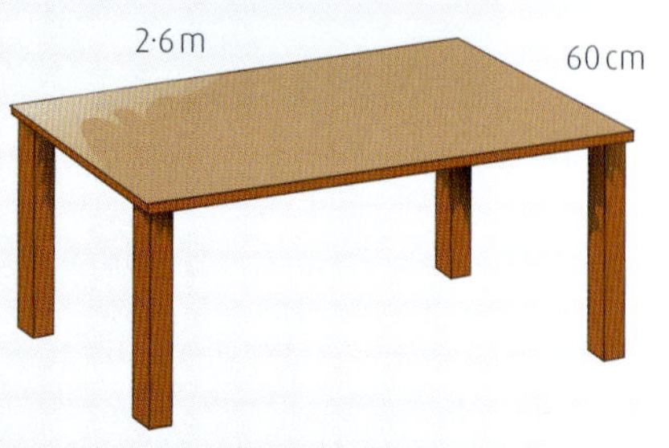

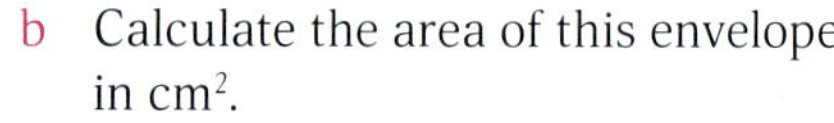

b Calculate the area of this envelope in cm².

c Calculate the area of this triangle

72 mm

20 cm

d Calculate the area of this path. Give your answer in square metres.

e The **straight part** of a running track is 84·39 metres long.
The width of each lane is 122 centimetres.
Calculate the area of the **straight part** of one lane.

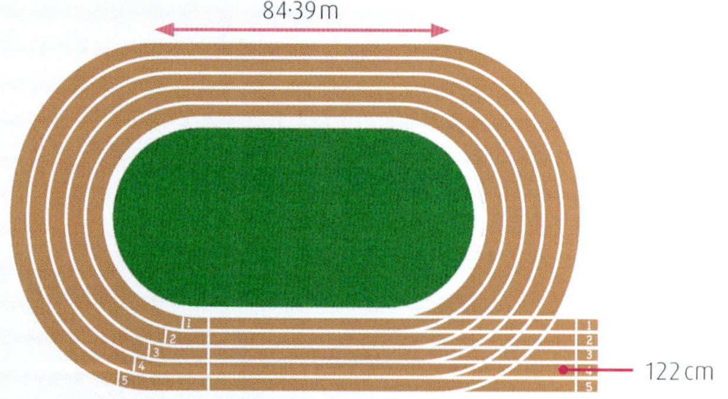

f Lead flashing, for putting on roofs come in lengths of 1·5 m
The flashing is 100 mm wide.
Calculate the area of one section of flashing.
Give your answer in square centimetres.

100 mm

g A rectangular shower door is 1850 mm high and 1 m wide.
What is the area of glass?
Give your answer in square centimetres.

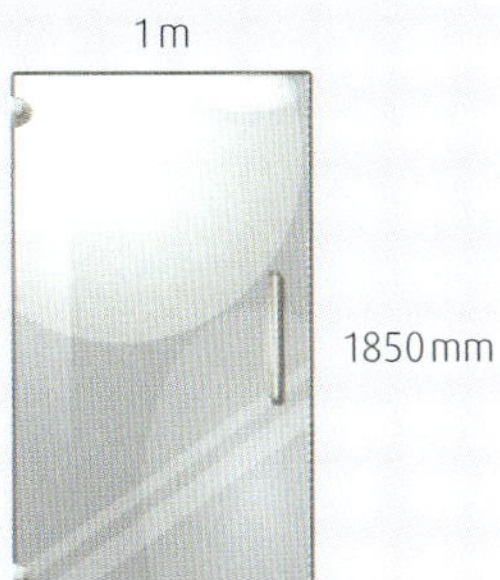

h 'The Eyes Have it' optician has a sign in the shape of a triangle as shown.
What is its area, in square centimetres?

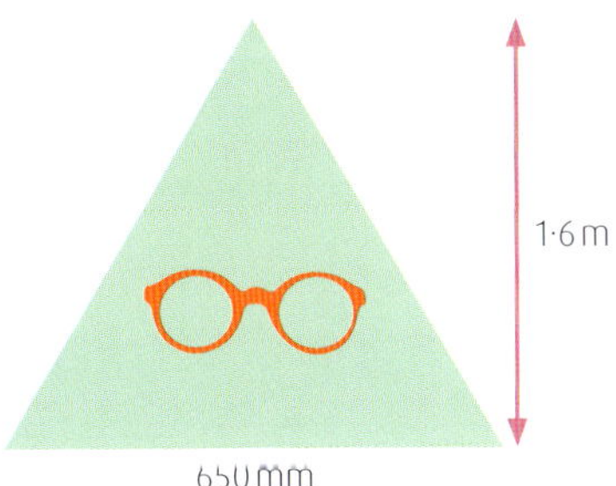

Volume of a cuboid

Volume is the amount of space inside a three-dimensional (3D) object.

Here are pictures of some cuboids in real life.

How does that work?

The diagram shows a cuboid.
This cuboid is made from 1 cm cubes.
There are 2 rows of 4 cubes on the base = 8 cubes.
There are 3 layers, giving 24 cubes.
The volume of this cuboid is found by doing:
$4 \times 2 \times 3 = 24$ cubic centimetres (cm^3).

In general, the volume of a cuboid is found by using the formula:

Volume = length × breadth × height
$V = lbh$

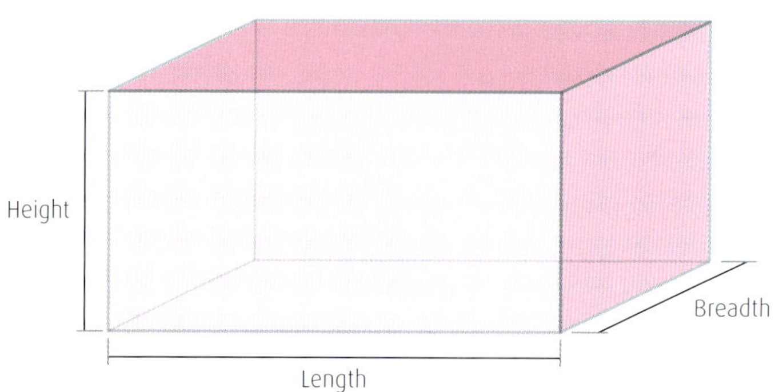

EXAMPLE

Calculate the volume of this biscuit tin.

SOLUTION

$V = lbh$
$= 30 \times 20 \times 15$
$= 9000\,\text{cm}^3$

DON'T FORGET

- Formula
- Substitute
- Calculate

Classroom challenge

1 Calculate the volume of these cuboids.

a

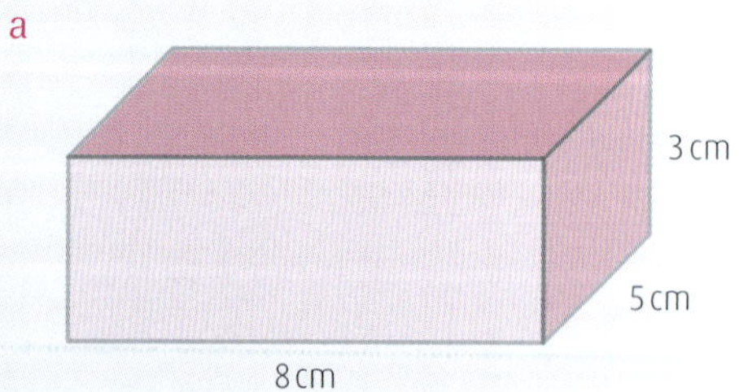

b

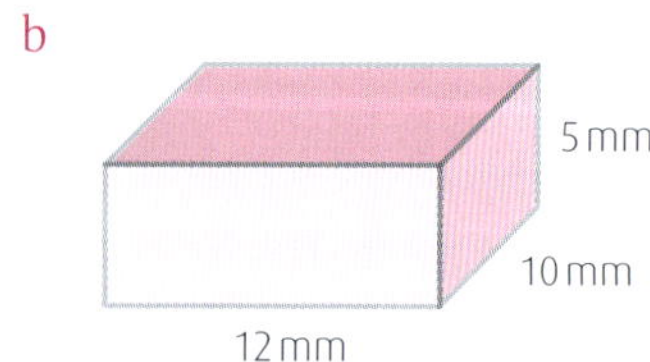

2 Calculate the volume of this cereal box.

3 Calculate the volume of this fish tank:
  a in cubic centimetres
  b in litres.

DON'T FORGET

1 litre = 10 000 cm^3

4 A standard brick is in the shape of a cuboid. Its dimensions are; length 15 cm, breadth 8 cm and height 5 cm. Calculate the volume of the brick.

5 A cuboid has length of 8 mm, breadth 6 mm and height 4 mm.
  a Calculate the volume of the cuboid.
  b A cube has a side of length 5 mm. Does the cube or the cuboid have the bigger volume?

6 A garden pond is a cuboid of length 8 m, breadth 4 m and depth 1·5 m. Calculate the volume of the pond.

7 A car has a fuel tank in the shape of a cuboid. The dimensions are; 50 cm by 60 cm by 20 cm.
  a Calculate the volume of the fuel tank in cubic centimetres.
  b Write the volume of the fuel tank in litres.

8 The picture shows a block of cheese. Calculate the volume of cheese.

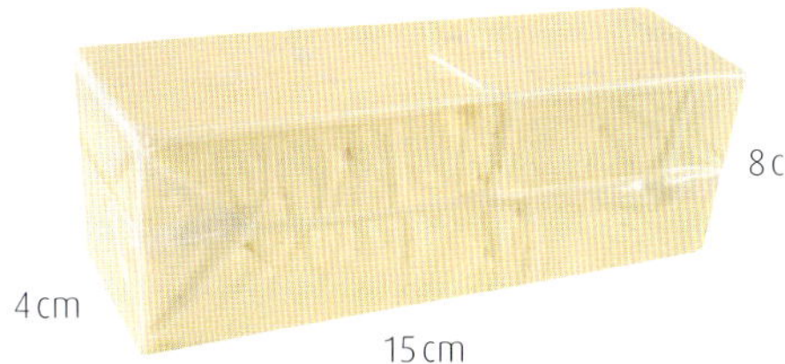

9 A carton of orange juice measures 9 cm by 6 cm by 19·5 cm. Show that its volume is about 1 litre.

10 A rectangular swimming pool is 25 m long and 12 m wide.
How many litres of water will it take to fill the pool to a depth of 1·8 m?

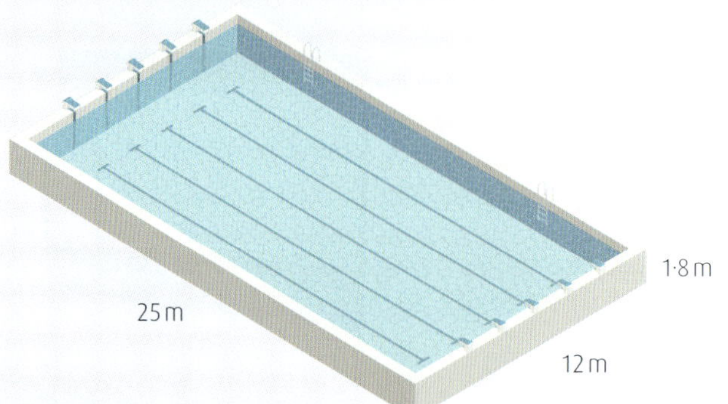

11 Du Rollie's is making ice cream.
Ice cream mixture is poured into a block measuring 20 cm by 7 cm by 5 cm.
  a Calculate the volume of one block.
  b How many blocks could be made from 4·2 litres of mixture?

12 The diagram shows a large container box and a smaller box.

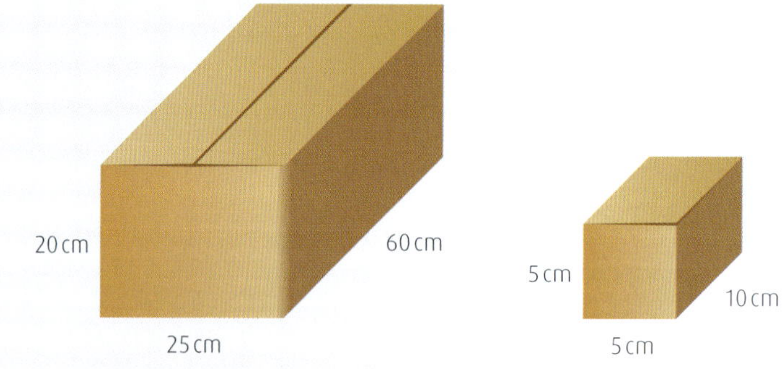

  a Calculate the volume of the large container box.
  b Calculate the volume of the smaller box.
  c How many smaller boxes will fit into the large container box?

Be careful – look at the units!

13 For each of these cuboids, calculate the volume.

a

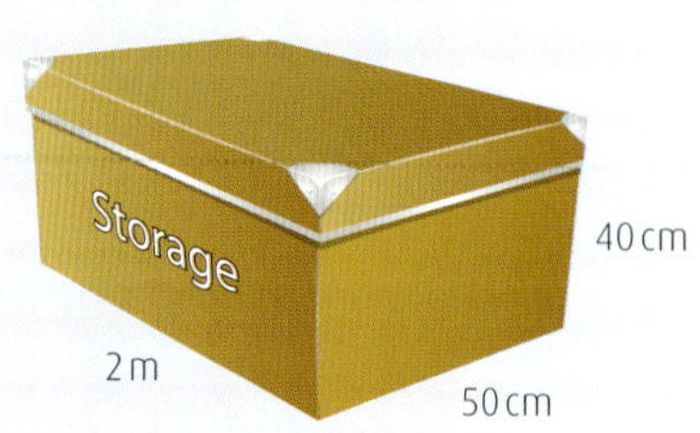

Calculate the volume in cubic centimetres.

b

Write your answer in cubic millimetres.

c

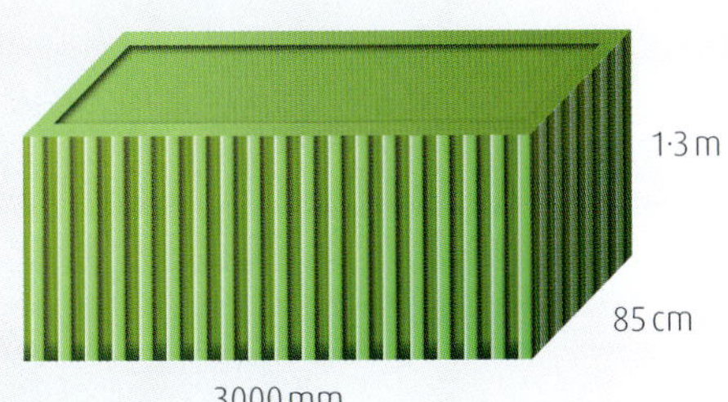

Calculate the volume in cubic metres.

d A concrete block is 2·5 metres long.
The width is 12 cm and height 10 cm.
Calculate the volume of the block in cubic centimetres.

e A builder is laying a tarmac drive in front of a new country house.
The drive is 11 metres long and 3 metres wide.
The tarmac needs to be 25 centimetres thick.
Calculate the volume of tarmac required.

f A rectangular paddling pool is 2·4 m long and 1·2 m wide.
Calculate how many litres of water are required to fill the pool to a depth of 30 cm.

g Calculate the volume of concrete needed to make a rectangular path which is 40 metres long, 80 centimetres wide and 150 millimetres thick.

STRETCH YOURSELF

14 Freya puts a rock in a tank as shown, and then fills the tank with water.
She takes the rock out and the water level drops by 5 centimetres.

a What is the volume of the empty tank? Give your answer in litres.
b What volume of water is left after the rock is taken out?
c Calculate the volume of the rock, in cm^3.
d How many litres of water would be needed to fill the tank up again?

Compound shapes and objects

Having investigated different routes to a solution, I can find the area of compound 2D shapes and the volume of compound 3D objects, applying my knowledge to solve practical problems. MTH 3-11b

What's coming up?

This Outcome and Experience will give you the opportunity to:

- find the area of compound 2D shapes constructed from squares, rectangles and triangles
- find the volume of compound 3D objects constructed from cubes and cuboids.

What you already know

You have already learned how to:

- ✔ calculate the area of squares, rectangles and right-angled triangles in square millimetres (mm^2), square centimetres (cm^2) and square metres (m^2)
- ✔ calculate the volume of cubes and cuboids in cubic centimetres (cm^3) and cubic metres (m^3).

Area of compound shapes

A compound shape is a shape made up of two or more 'basic' shapes.

The area is found by calculating the area of each individual shape and then adding them together.

How does that work?

EXAMPLE

Calculate the area of this shape.

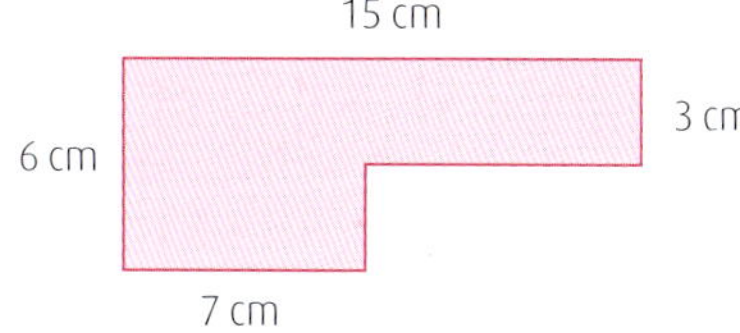

SOLUTION

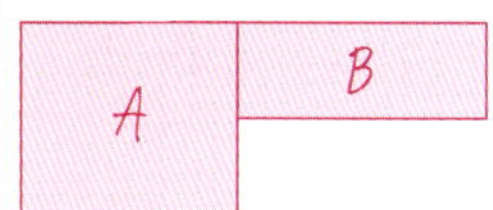

Step 1

Divide shape into basic shapes you know.
In this case, two rectangles.

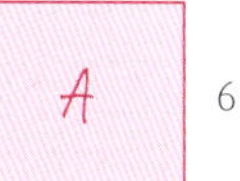

Step 2

Calculate each area

Area A = lb
$= 7 \times 6$
$= 42\,\text{cm}^2$

Area B = lb
$= 8 \times 3$
$= 24\,\text{cm}^2$

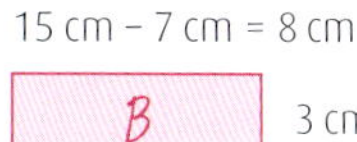

Step 3

Calculate total

The total area of the shape
$= 42 + 24 = 46\,\text{cm}^2$

EXAMPLE

Calculate the area of this shape.

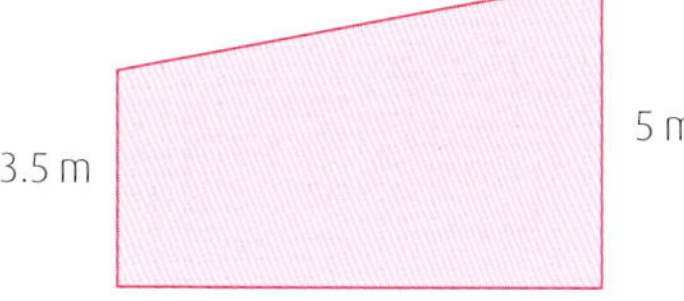

SOLUTION

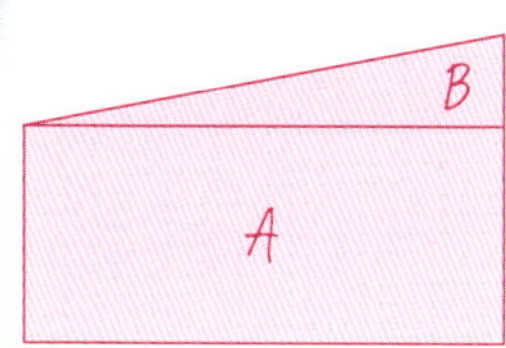

Step 1

Divide the shape into basic shapes you know.
In this case, a rectangle and a triangle.

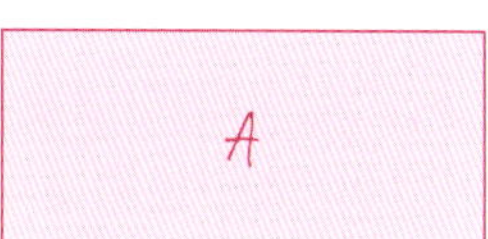

Step 2

Calculate each area.

Area A = lb
$= 8 \times 3{\cdot}5$
$= 28\,\text{m}^2$

Area B = $\frac{1}{2}bh$
$= \frac{1}{2} \times 8 \times 1{\cdot}5$ ($5\,\text{m} - 3{\cdot}5\,\text{m} = 1{\cdot}5\,\text{m}$)
$= 6\,\text{m}^2$

Step 3

Calculate total area.

The total area of the shape
$= 28 + 6 = 34\,\text{m}^2$

EXAMPLE

Calculate the shaded area.

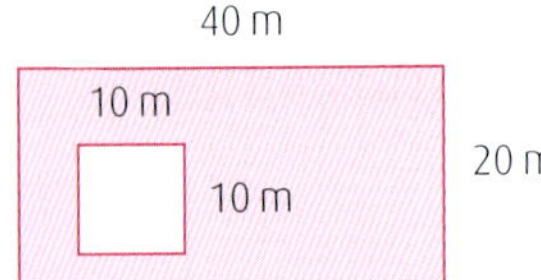

SOLUTION

Step 1
Divide the shape into basic shapes you know.
In this example, a rectangle and a square.

Step 2
Calculate each area.

Rectangle

Area = lb
= 40×20
= $800\,\text{mm}^2$

Square

Area = l^2
= 10^2
= $100\,\text{mm}^2$

Step 3
Calculate shaded area
The shaded area
= rectangle − square
= 800 − 100
= $700\,\text{mm}^2$

Classroom challenge

1 Calculate the area of each of these compound shapes.

a

6 cm
3 cm
4 cm
10 cm
7 cm
2 cm

b

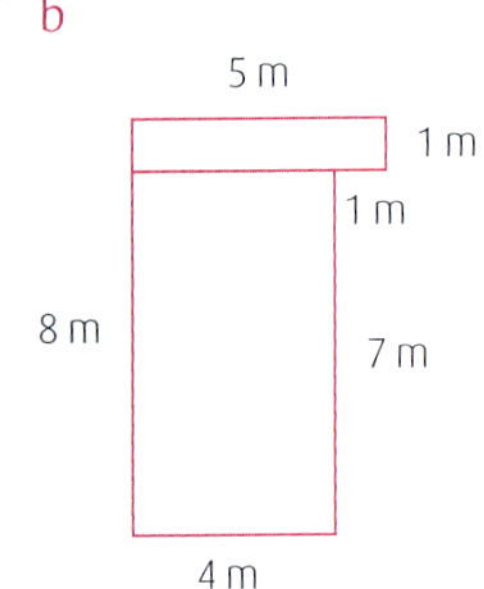

2 Calculate the area of each of these compound shapes.
In these, you must decide on how you will split up the shapes.

a

4 m
6 m
8 m
5 m

b

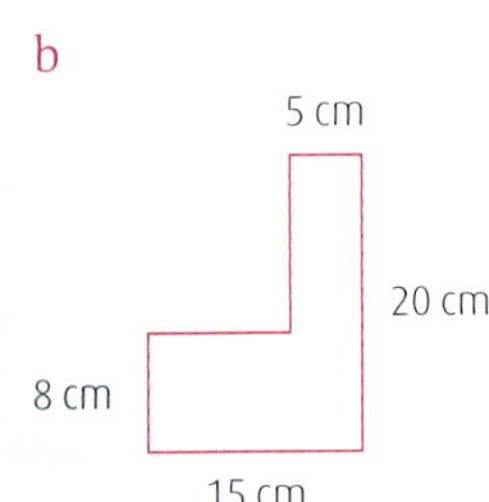

c

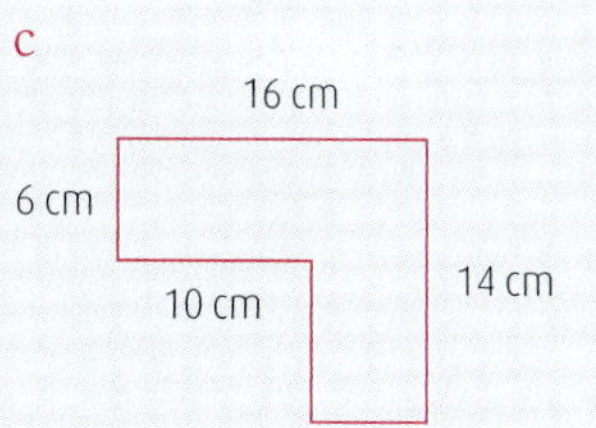

d

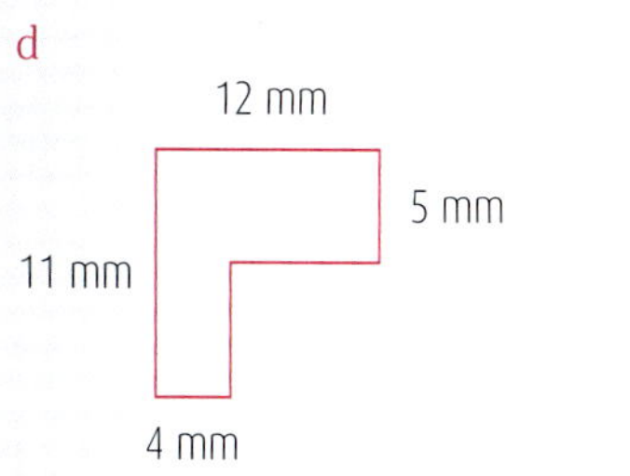

3 Calculate the total area of each of these shapes.

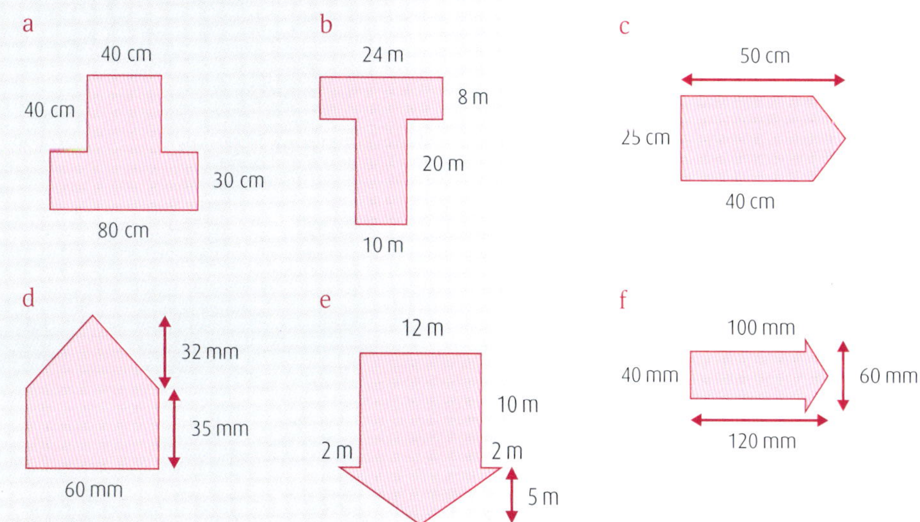

4 Calculate the total area of these shapes.

a

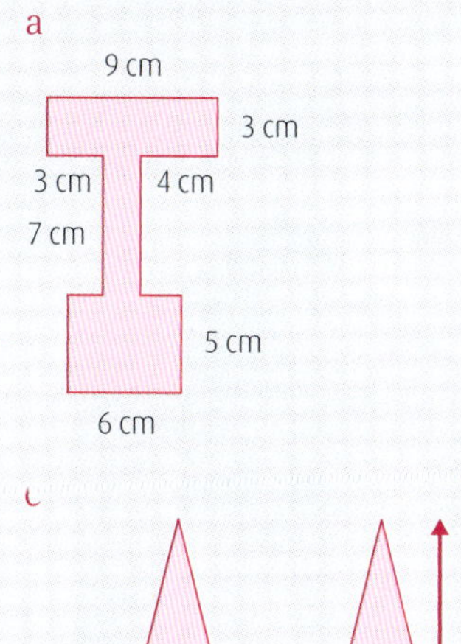

b

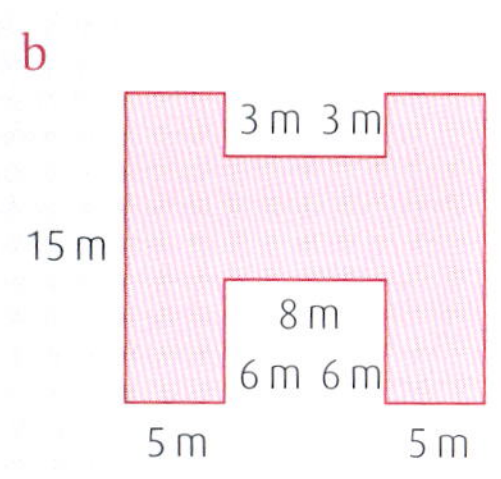

c

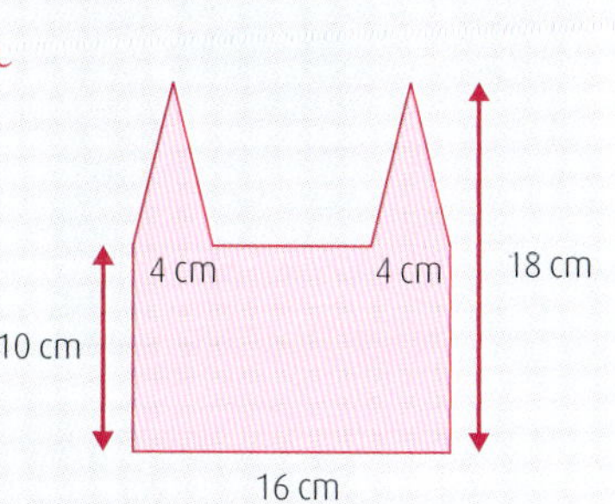

d

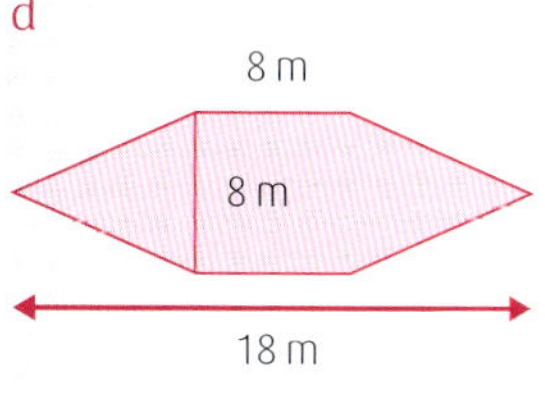

5 Calculate the shaded area in each of these shapes.

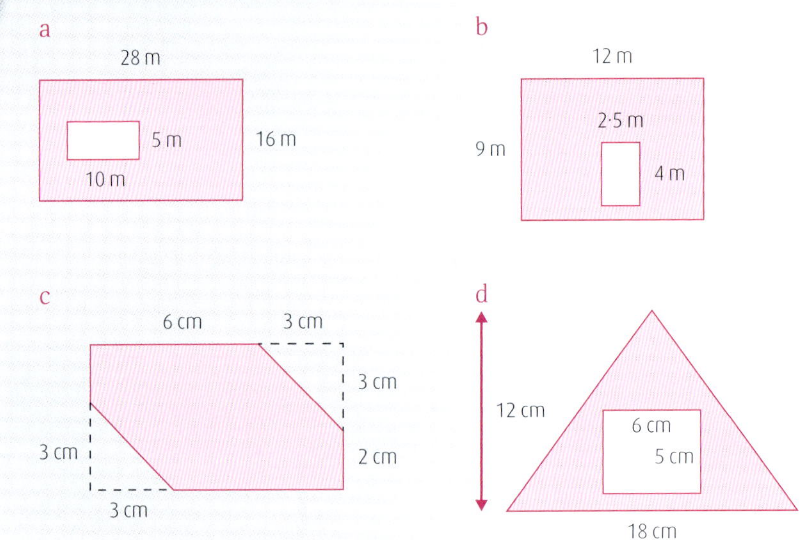

STRETCH YOURSELF

6 Janine is going to paint the front of her house.

a Find the total area to be painted.

b A 5-litre tin of masonry paint costs £25.
1 litre of masonry paint covers 2 m^2 of wall.
Janine has set aside £300.
Does she have enough to paint her house?
Show all your working to justify your answer.

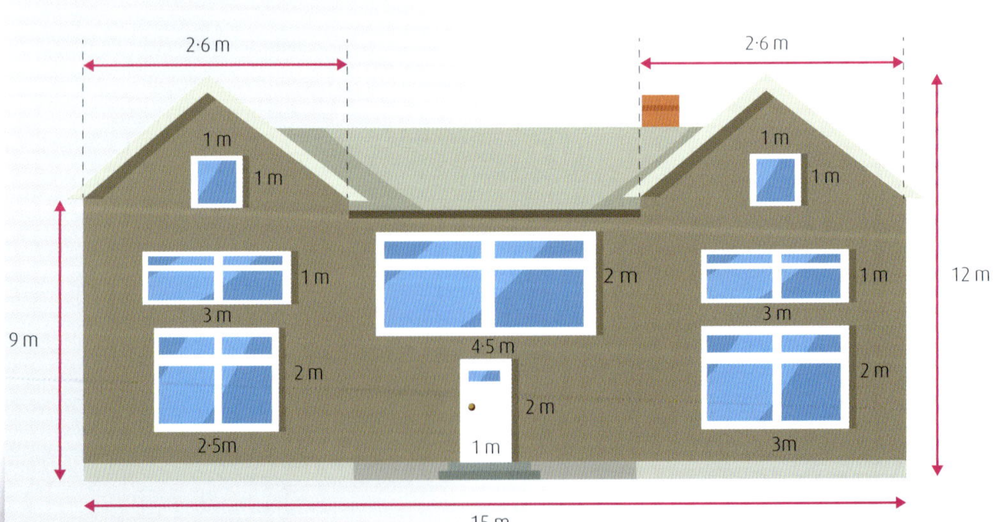

Volume of compound objects

You can calculate the volume of compound composite volumes in the same way as you calculate the area of shapes.

How does that work?

Use the same three steps as for finding areas.
Here is an 'L' shaped object.
It is made from two cuboids.
Following the steps above:

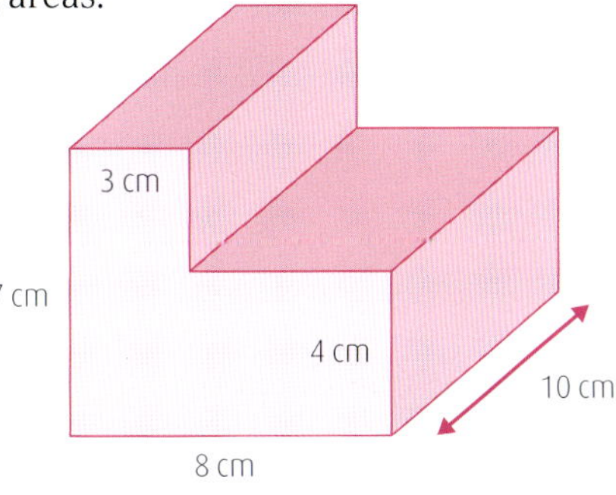

Step 1

Divide the object into basic 3D objects (such as cubes or cuboids).

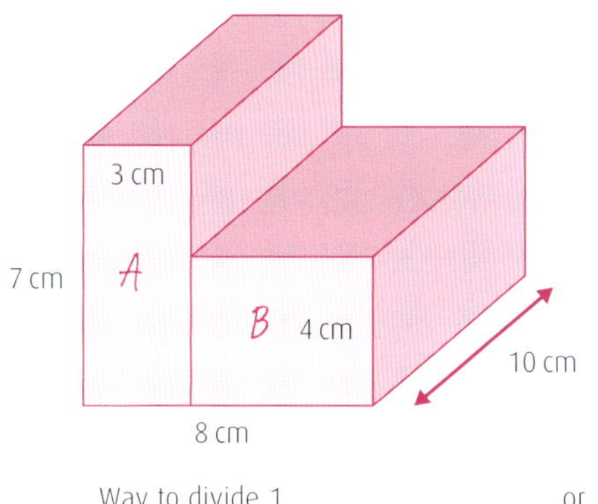

Way to divide 1

or

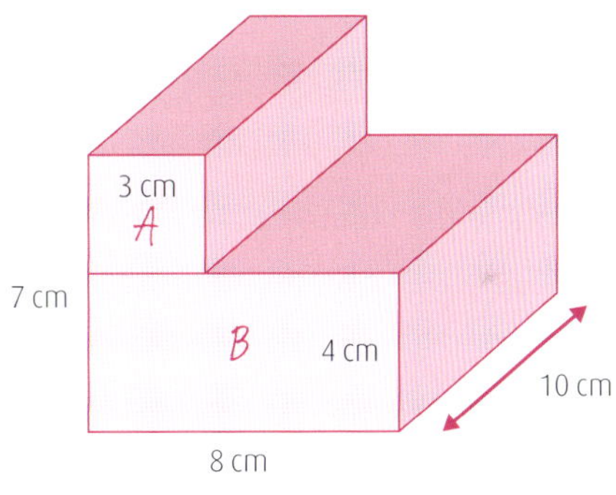

Way to divide 2

Step 2

Calculate the volume of each object.

Diagram 1

Volume of A
V $= lbh$
$= 7 \times 3 \times 10$
$= 210\,\text{cm}^3$

Volume of B
V $= lbh$
$= 4 \times \mathbf{5} \times 10$
$(8 - 3 = 5)$
$= 200\,\text{cm}^3$

Step 3

Calculate the total.

Total volume $= 210 + 200$
$= 410\,\text{cm}^3$

Diagram 2

Volume of A
V $= lbh$
$= \mathbf{3} \times 3 \times 10$
$(7 - 4 = 3)$
$= 90\,\text{cm}^3$

Volume of B
V $= lbh$
$= 8 \times 4 \times 10$
$= 320\,\text{cm}^3$

Total volume $= 90 + 320$
$= 410\,\text{cm}^3$

Classroom challenge

1 Calculate the volume of this 'L' shaped object.

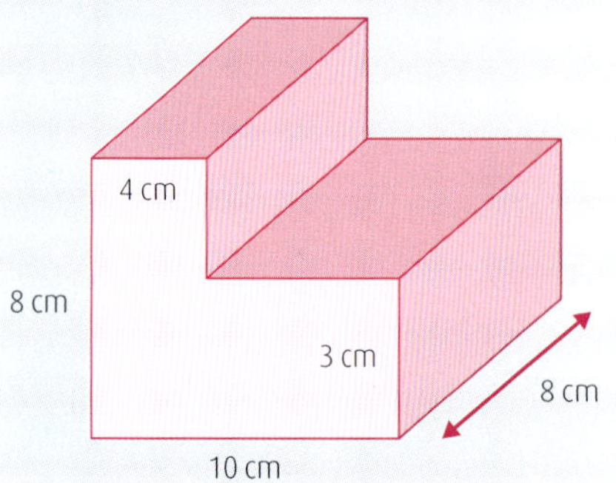

2 Calculate the volume of this solid.

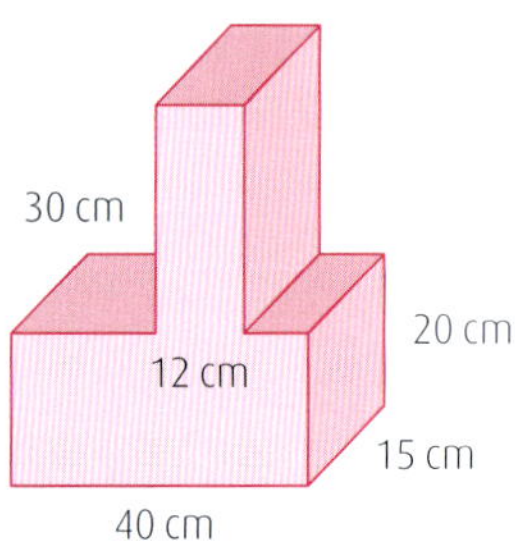

3 Calculate the volume of this object.

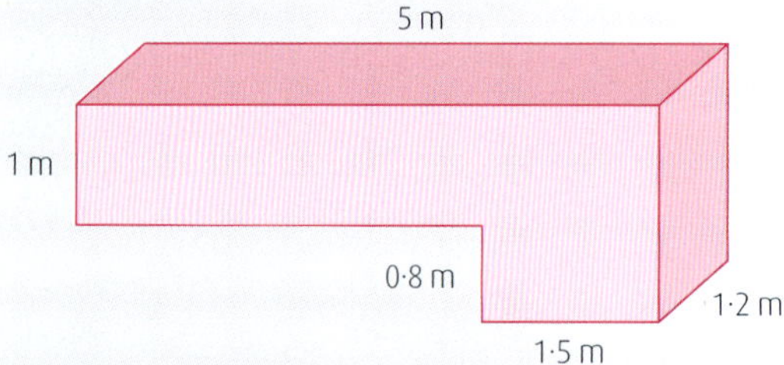

4 Calculate the volume of this object.

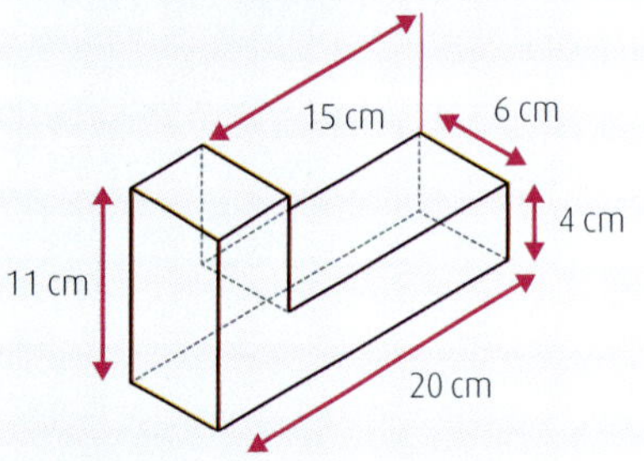

5 Calculate the volume of this 'U' shaped object.

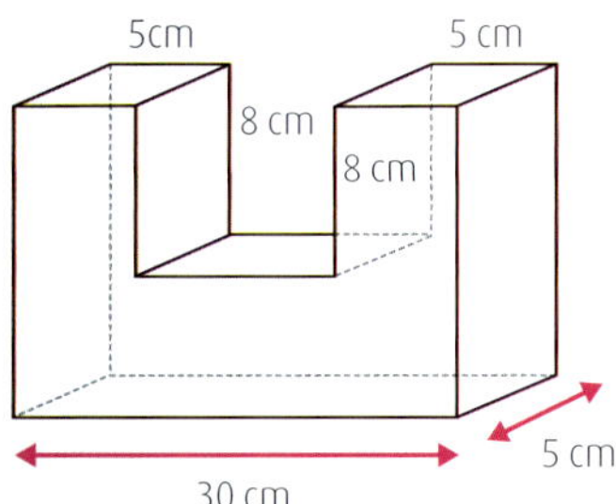

6 A children's foam climbing tunnel is made in the shape of a cuboid, with a cuboid tunnel cut out of the centre. What is the volume of foam used to make the tunnel?

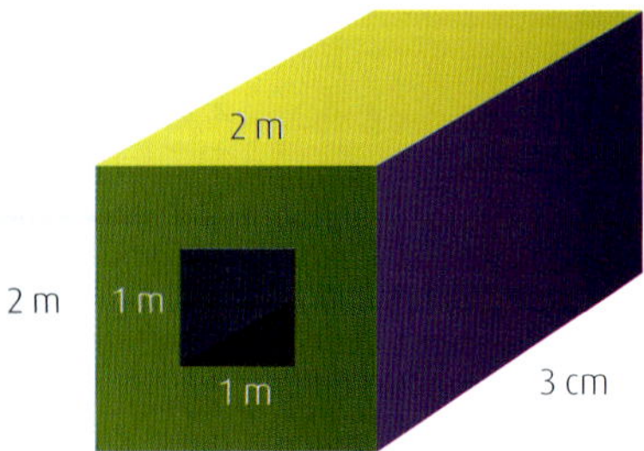

7 The diagram shows a wooden block that has had a square hole cut out of it. Calculate the volume of wood in the block.

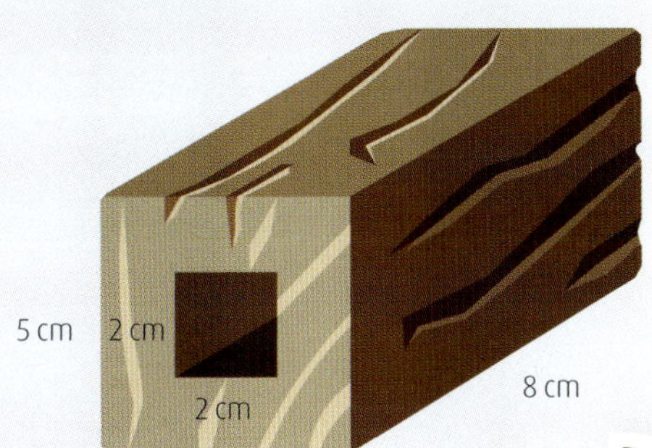

8 Calculate the volume of this garden wall.

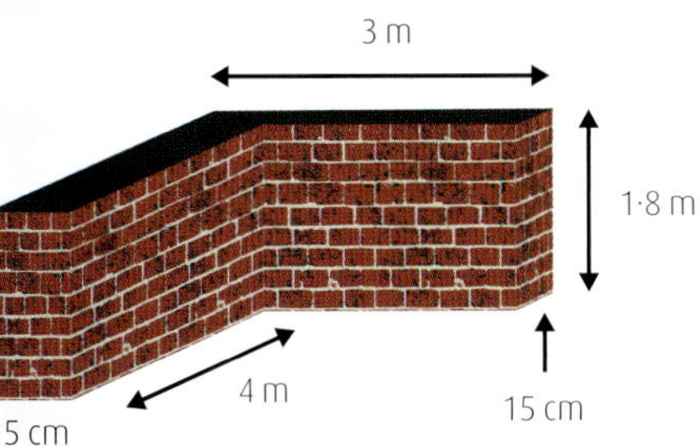

9 The picture and diagram show three wooden 'sleepers' which form steps up to a garden. What is the total volume of the sleepers?

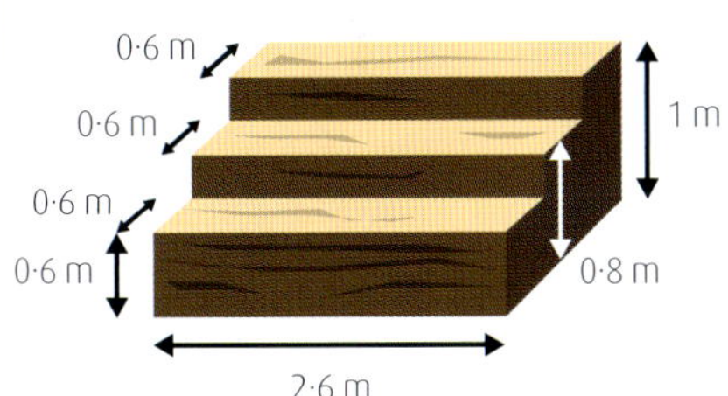

10 The diagram shows a sculpture outside Hubert's factory. The sculpture is made of brass. What is the volume of brass in the sculpture?

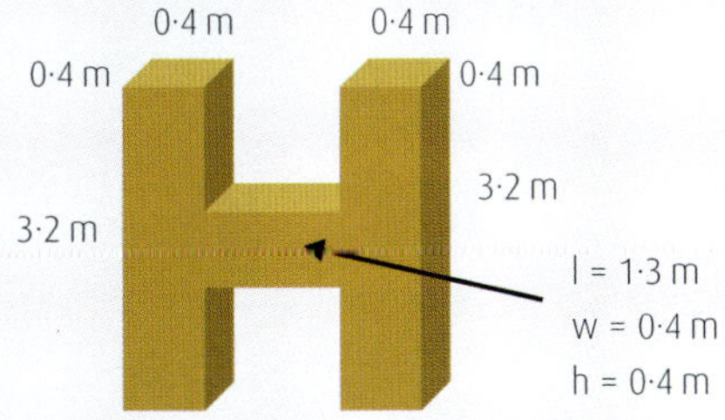

11 The sketch shows an 'impossible triangle'. It looks as though it consists of identical cuboids. Calculate the volume of this 'impossible triangle'.

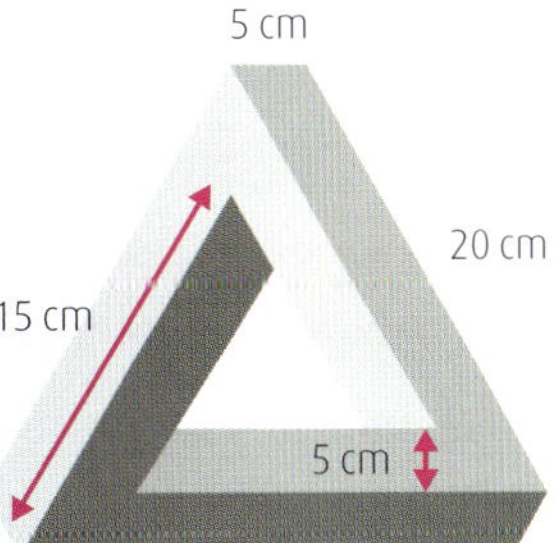

IMPACT OF MATHEMATICS

RESEARCH AND PROJECT WORK

I have worked with others to research a famous mathematician and the work they are known for, or investigated a mathematical topic, and have prepared and delivered a short presentation. MTH 3-12a

What's coming up?

This Outcome and Experience will give you the opportunity to:

- research and communicate using appropriate mathematical vocabulary and notation, the work of a famous mathematician or a mathematical topic and explain the relevance and impact they have on society.

What you already know

You have already learned how to:

- ✔ research and present examples of the impact mathematics has in the world of life and work
- ✔ contribute to discussions and activities on the role of mathematics in the creation of important inventions, now and in the past.

Learning and discovering mathematics is more than simply learning rules, strategies, calculations and how to solve problems.

A portrait of Pythagoras (*Pythagoras' Head*, Domenico Cunego).

Mathematics is best learned when you have a 'feel' for where it has come from, who has been involved, how it has changed everyday life and how it improves our lives; or in other words – the impact it has on all of us.

How does that work?

In this section, it is suggested that you choose a mathematician, or a mathematical topic.

Then, as a small group, research your choice and present your findings to the class.

There are a couple of suggestions below to help you get started.

A portrait of John Napier

Famous mathematician

Choose a mathematician to study.

- They could be from a long time ago, for example, Pythagoras, Hypatia, Euclid.
- A bit closer to now, for example, John Napier, Isaac Newton, Gottfried Leibniz.
- Or more up to date, for example, Alan Turin, Andrew Wiles, Albert Einstein.
- Research or find out some facts about him/her.
- When were they born?
- Where were they born; where did they live?
- Was mathematics their 'job' or did they do another job?
- What were conditions like at that time?
- Were there other 'famous' mathematicians working on similar ideas?
- What piece of mathematics are they known for?
- How has this shaped our world today?

Put this into a presentation:

- a poster
- a 'Powerpoint' show
- a video
- a talk.

Andrew Wiles

A mathematical topic

Choose a topic. For example, Pythagoras' theorem, the Fibonacci sequence, algebra.

- Find out who discovered or developed the topic.
- When was it discovered or developed?
- How long did it take for the topic to appear in 'everyday life'?
- What impact did it have on the world when it was discovered?
- What impact does it still have today?
- Did other mathematicians work on similar ideas?
- Was the discovery made in war time or peace time?

Put this into a presentation:

- a poster
- a 'Powerpoint' show
- a video
- a talk.

Muhammad ibn Mūsā al-Khwārizmī, who used 'al-jabr' in the title of his book.

Fibonacci: Mathematics in nature

- Research the Fibonacci sequence and where it appears in nature.
- How is Fibonacci sequence used elsewhere, for example to convert between miles and kilometres?
    - Look at this sequence: 1, 2, 3, 5, 8, 13, 21, 34, 55, 89
    - 8 kilometres is about 5 miles. Find 8 in the sequence, now look to the left!
    - 34 miles is about 55 kilometres. Find 34 in the sequence, now look to the right!

- Is Fibonacci used elsewhere, for example in security?
- Here are some interesting areas in nature that you may wish to investigate further.
- A tree growing – how do branches form and split?

The main trunk grows and produces 1 branch.
Then one of the stems branch into 2, whilst other remains.

This pattern of branching repeats for each new stem.
Look at trees in a local park – can you see the pattern?
Flowers: let's look at petals!
Surprisingly, the numbers of petals on flowers follow the Fibonacci sequence.

Number of petals	Example of flower	Picture
3	Clover	
5	Buttercup	
8	Violet	

Number of petals	Example of flower	Picture
13	Black eyed Susan	
21	Chicory	
34	Daisy	
Can you find more?		

Others: look at pictures of

- Sunflowers
- pine cones
- shells
- think big ... look at galaxies
- think wild ... look at satellite pictures of hurricanes.

Present your findings to your class.

Numbers!

You may wish to investigate numbers and present types of numbers to the class – and their use in life today.

Some numbers you may wish to investigate might include:

- square numbers
- triangular numbers
- prime numbers (investigate internet security)
- perfect numbers
- irrational numbers
- binary numbers (computers)
- exponential numbers
- Roman numerals
- Egyptian numbers.

There are 10 types of people in the world – those who know binary, and those who don't

Other ideas might include:

- When were counting numbers first used?
- When was '0' first used?

Present your findings to your class.

Marin Mersenne

Hieroglyphic numbers

ALGEBRAIC SKILLS

PATTERNS AND RELATIONSHIPS

Having explored number sequences, I can establish the set of numbers generated by a given rule and determine a rule for a given sequence, expressing it using appropriate notation. MTH 3-13a

What's coming up?

This Outcome and Experience will give you the opportunity to:

- generate number sequences from a given rule, for example, $t = 4n + 6$
- extend a given pattern and describe the rule
- express sequence rules in algebraic notation, for example, the cost of hiring a car is £75 plus a charge of £0·05 per mile, 'm' driven, $C = 0{\cdot}05m + 75$.

What you already know

You have already learned how to:

✔ explain and use a rule to extend well known number sequences including square numbers, triangular numbers and Fibonacci sequence

✔ apply knowledge of multiples, square numbers and triangular numbers to generate number patterns.

Patterns and sequences

Some number patterns are familiar, and some may take a little more thought!

How does that work?

A familiar number pattern is 2, 4, 6, 8, 10, 12, … .
A 'rule' for this pattern could be 'even numbers starting at 2'
or it could be 'start at 2 and go up in twos'.

Another familiar pattern is 1, 3, 5, 7, 9, 11, … .
A 'rule' for this pattern could be 'odd numbers starting at 1'
or it could be 'start at 1 and go up in twos'.

A **sequence** has a **starting point** and a **rule** to find the **next term**.
The starting point is usually called the **1st term**.
Each **term** in a **sequence** is made by following the **same rule**.

EXAMPLE

Starting point 4
Rule add 3 (+3)
Sequence 4, 7, 10, 13, 16, 19, … …

EXAMPLE

1st term 1
Rule double (×2)
Sequence 1, 2, 4, 8, 16, 32, … …

EXAMPLE

1st term 60
Rule subtract 5 (– 5)
Sequence 60, 55, 50, 45, 40, 35, … …

Classroom challenge

1 For each row in the table, create a sequence using the 1st term and the given rule.
For each one, write down the first 6 terms.
The first one has been done for you.

1st term	Rule	Sequence
1	Add 3	1, 4, 7, 10, 13, 16, … …
3	Double	
4	Add 6	
2	Multiply by 5	
80	Subtract 5	
10	Multiply by 10	
6	Add 6	

2 Write down the next 2 terms for each of these sequences.
Also write down the rule you used to get these terms.
a 2, 6, 10, 14, 18, 22, …, …
b 7, 15, 23, 31, 39, …, …
c 90, 87, 84, 81, 78, …, …
d 4, 12, 36, 108, 324, …, …
e 88, 77, 66, 55, 44, …, …
f 1, 5, 25, 125, 625, …, …
g 81, 72, 63, 54, …, …

3 Write down the next 2 terms of these sequences.
Write down the rule you used to get these terms.
a 5, 10, 15, 20, …, …
b 4, 8, 12, 16, …, …
c 9, 18, 27, 36, …, …
d 10, 20, 30, 40, …, …
e 3, 6, 9, 12, …, …
f 2, 4, 6, 8, 10, …, …
g 7, 14, 21, 28, …, …

Look at question 3a above.

Term	1	2	3	4	5	6
Value	5	10	15	20	25	30

5 5 5 5 5

The values go up in 5s. Exactly like the 5 times table!

5 × **1** = 5 gives **1**st term
5 × **2** = 10 gives **2**nd term
5 × **3** = 15 gives **3**rd term
5 × **4** = 20 gives **4**th term

This suggests that the **5**th term will be found by doing
5 × **5** = 25 which it is!
This suggests that the **6**th term will be found by doing
5 × **6** = 30 which it is!
It also shows a rule to find any term.
For example, to find the **100**th term simply multiply by 5
5 × **100** = 500
So, to find the '***n***th' term simply multiply by 5
5 × ***n*** = 5***n***
This gives us a rule for finding any term in the sequence
nth term = 5***n***

4 a What will the 7th term be?
b What will the 10th term be?
c What will the 1000th term be?

5 Copy and complete this table for the sequence in question 3b above.

Term	1	2	3	4	5	6
Value	4	8	12	16		

a What will the 7th term be?
b What will the 10th term be?
c What will the 100th term be?
d What will the *n*th term be?

6 Repeat question 5 for the sequences in question 3c to 3g.
Complete the rules for finding the '***n***th' term.
a *n*th term = × *n* =
b *n*th term = × *n* =
c *n*th term = × *n*
d *n*th term
e *n*th term

7 Use the rule given to write down the first 5 terms of each sequence.
The first one has been done for you.
a $3n + 2$ $n = 1$, gives $3 \times 1 + 2 = 5$, $n = 2$, gives $3 \times 2 + 2 = 8$, $n = 3$ gives $3 \times 3 + 3 = 11$
$n = 4$, gives $3 \times 4 + 2 = 14$, $n = 5$ gives $3 \times 5 + 2 = 17$
First 5 terms are 5, 8, 11, 14, 17

b $4n + 1$
c $5n + 3$
d $6n - 1$
e $7n$
f $8n - 3$
g $10n - 2$
h $12n$
i $11n$
j $3n - 2$

Sequences and algebraic notation

Sometimes a sequence may be generated by using a rule or a formula.

How does that work?

John is a plumber.
When he is called out on an emergency he charges a £50 call-out fee plus £20 per hour for the time taken to fix the problem.
Construct a formula to show John's charge.
Let's call the charge C and the number of hours taken to be h.
The charge is made up of a fixed fee of £50 and £20 × the number of hours

That is $C = £50 + £20 \times h$

The formula for calculating the charge is $C = 50 + 20h$

This can now be used to calculate a charge for any number of hours taken.

Number of hours (h)	1	2	3	4	5	6	7
Calculation of charge	50 + 20 × 1	50 + 20 × 2	50 + 20 × 3	50 + 20 × 4	50 + 20 × 5	50 + 20 × 6	50 + 20 × 7
'Sequence' £	70	90	110	130	150	170	190

If John spent 10 hours on an emergency repair, how much should he charge?

Formula $C = 50 + 20h$
Substitute $= 50 + 20 \times 10$
Calculate $= 50 + 200$
$C = 250$
John should charge £250 for the job.

Classroom challenge

1 Kelly is an emergency engineer.
She charges a call-out fee of £60 plus £30 per hour when doing an emergency repair.

a Construct a formula for Kelly's charge.
Let the charge be C and the number of hours worked be h.

Copy and complete:
The charge is made up of a fixed fee of and × number of hours
$C =$ £ + £ ×
$C =$

b Copy and complete this table to show Kelly's charge for different numbers of hours.

Number of hours (h)	1	2	3	4	5	6	7
Calculation of charge	60 + 30 × 1						
'Sequence' £	90						

2 A window cleaner charges £2 to visit a house and then £0·50 for every window cleaned.
  a Write down a formula for the window cleaner's total charge. Use C for charge and w for number of windows.
  b How much would the window cleaner charge if he visited a house with:
    i 5 windows
    ii 10 windows
    iii 15 windows?

3 A taxi driver charges a flat fee of £4 plus £1·20 per mile travelled.
  a Write down a formula for the taxi driver's charge. Use C for charge and m for miles.
  b Jane takes the taxi from work to home, a distance of 8 miles. How much would she be charged?

4 Neptune chocolate bars cost 32p each.
Write down a formula for the cost, C pence, of n Neptune bars.

5 Petrol costs £1·33 per litre.
Write down a formula for the cost, C pounds, of l litres of petrol.

6 A maths tutor charges £25 per hour for lessons and a fixed fee of £12 for materials.
  a How much would she charge for 3 lessons?
  b Write down a formula to calculate the cost of l lessons. Use $C = ...$
  c How much would she charge for (i) 5, (ii) 10, (iii) 15 lessons?

7 Dave and Carol saw this advert for hiring a narrowboat.
  a Write down a formula for the cost (C) of hiring the narrowboat if you know the number of days d.
  b Copy and complete this table for costs for different number of days hire.

Number of days (d)	1	2	3	4	5	6	7
Calculation of charge	100 + 45 × 1						
'Sequence' £	90						

  c How much would it cost Dave and Carol to hire the narrowboat for 12 days?

8 Bert's Bicycle holidays charges £65 per day plus a fixed charge of £25 for the hire of a bike and a helmet.
  a Make up a table to show the number of days (d) and the total cost (C) for up to 5 days hire.
  b Write down a formula for calculating the cost of d days hire.
  c How much would it cost to hire a bike from Bert's for (i) 7 days (ii) 14 days?

9 Azim is an electrician. He charges a call-out fee of £50 and £85 per hour.
  a Construct a formula for Azim's charge.
  b How much would Azim charge for 5 hours work?

STRETCH YOURSELF

  c If Azim charged a customer £390, how many hours had he worked?

EXPRESSIONS AND EQUATIONS

Expressions

I can collect like algebraic terms, simplify expressions and evaluate using substitution.
MTH 3-14a

What's coming up?

This Outcome and Experience will give you the opportunity to:
- collect like terms, including squared terms, to simplify an algebraic expression
- evaluate expressions involving two variables using both positive and negative numbers.

What you already know

You have already learned how to:
- ✓ apply the correct order of operations in number calculations when solving multi-step problems
- ✓ work with negative numbers.

Simplifying expressions

When working with algebra, you will need to understand some key words that describe what you are doing.

Word	Meaning
Variable	This is a letter in a term, or expression, which can take different values, that is, it can vary. In algebra we can use any letter as a variable, but the most common ones are x, y, n, t, w.
Operator	An operator is the 'action' you do: add, subtract, multiply, divide. An operator would go between terms in an expression. For example, $3x + 4$, $2t - t + 4$, and so on. When using variables, the times sign '×' is missed out (as it may get confused with an 'x'). For example, $3 \times x$ is written as $3x$, $t \times 5$ is written as $5t$, $t \div 2 = \frac{t}{2}$, and $2 \div t = \frac{2}{t}$ and so on.
Term	A term is one item in an expression. A term can contain a letter, a number or both. For example; $3n$ is a term, W^2 is a term, $\frac{n}{3}$ is a term. $3n + 4$ has two terms: $3n$ and 4.
Expression	An expression is a collection of terms, with operators between them. For example. $n + 8$ is an expression that means 'add 8 to n'. $6 - w$ is an expression that means 'subtract w from 6'. $3x + 4$ is an expression that means 'multiply x by 3 then add 4'. $2x + 3y - 4$ is an expression that means 'multiply x by 2 then add 3 times y, then subtract 4'.
Coefficient	A coefficient is the number which goes with the variable in a term. For example, in $4x$ x is the variable and 4 is the coefficient. For example, in $-6y$ y is the variable and -6 is the coefficient.

If you think of mathematics as a language, there are 'rules' for how to write terms and expressions.

- If there is a number and a letter, the number comes first: $t \times 3$ is written as $3t$.
- If there is more than one letter, they are, usually, written in alphabetical order: $a \times d = ad$, $c \times 5d = 5cd$.
- The operator goes with what comes after it. $6p - 3p$ means 'start with $6p$ then subtract $3p$'.
- $1 \times n$ is written as n.
- $0 \times n$ is written as 0.
- $n \times n$ is written as n^2.

EXAMPLE

Simplify:

i $k + k + k$ ii $4a \times b$ iii $6p \times 3$ iv $2m \times 7n$ v $6x \div 2$

SOLUTION

i	$k + k + k$	$= 3 \times k = 3k$	
ii	$4a \times b$	leave out '×' sign and write letters in alphabetical order	$= 4ab$
iii	$6p \times 3$	multiply numbers together	$= 18p$
iv	$2m \times 7n$	multiply numbers together, and letters, leave out '×'	$= 14mn$
v	$6x \div 2$	divide 6 by 2, $= \frac{6x}{2}$	$= 3x$

Classroom challenge

1 Simplify each of these expressions.

a $t + t$
b $t + t + t + t$
c $d + d + d$
d $k + k + k + k + k$
e $a + a + a + a + a + a$
f $x + x + x$
g $g + g + g + g - g$
h $x + x + x - x$
i $m + m - m - m$
j $t + t + t + t + t - t - t$
k $n - n + n + n$
l $w + w - w$

2 Copy and complete this table.
The first one has been done for you.

a	$t + t + t + t + t$	$= 5 \times t$	$= 5t$
b	$m + m + m$		
c	$w + w + w + w + w + w + w + w$		
d		$= 7 \times x$	
e			$= 6k$
f			$= 10p$
g		$= 9 \times y$	

3 Write each of these expressions in a simpler form.

a	$4 \times m$	e	$k \times 3$	i	$n \div 3$	m	$t \times q$
b	$6 \times k$	f	$z \times 6$	j	$t \div 6$	n	$w \times w$
c	$9 \times t$	g	$y \times x$	k	$7 \div w$	o	$s \times s$
d	$2 \times r$	h	$b \times a$	l	$6 \div m$	p	$a \times a$

4 Simplify each of the following expressions.

a	$b \times 3a$	c	$3p \times 2q$	e	$2m \times 3m$	g	$3 \times 4m \times 2n$
b	$4m \times n$	d	$5d \times 3c$	f	$4k \times 3k$	h	$2a \times 3b \times 4c$

Collecting *like* terms

Like terms are multiples of the same letter, or the same combination of letters.
Like terms can also be letters to the same power.

How does that work?

$x, 3x, \frac{1}{2}x, -5x$	These are ***like terms***	they are all ***multiples*** of x.
$4pq, 7pq, -\frac{1}{2}pq, -2pq$	These are ***like terms***	they are all ***multiples*** of pq.
$w^2, 6w^2, \frac{1}{2}w^2, -3w^2$	These are ***like terms***	they are all ***multiples*** of w^2.

Only like terms can be added or subtracted – or 'collected together' – to simplify an expression.

For example, $4m + 3m$ simplifies to $7m$

$6k + 3k - 2k$ simplifies to $7k$ $(6 + 3 - 2 = 7)$

$4w^2 + 5w^2$ simplifies to $9w^2$

Where there is more than one variable in the expression, you can only collect together **like terms**.

For example, $3m + 6m + 7n - 2n$

Collect together like terms $(3m + 6m) + (7n - 2n)$

This simplifies to $9m + 5n$

When simplifying an expression, follow these steps.

Step 1 Rearrange so the same letters are 'grouped' together.

Step 2 Simplify each group.

EXAMPLE

Simplify $4w + 3t + 6w - 2t$

Step 1

Rearrange $4w + 6w + 3t - 2t$

Step 2

Simplify each group $10w + t$

Note: do not write '$1t$', $1t = t$; $1g = g$; $1x = x$.

DON'T FORGET

Remember: the sign goes with what comes after it.

EXAMPLE

Simplify $3a + 6b + 5a + 5b + c + 3c - 2b - 2a$

Step 1 $3a + 5a - 2a \quad + \quad 6b + 5b - 2b \quad + \quad c + 3c$

Step 2 $6a \quad + \quad 9b \quad + \quad 4c$

$3a + 6b + 5a + 5b + c + 3c - 2b - 2a = \quad 6a + 9b + 4c$

Classroom challenge

1 Simplify each of these expressions.

a $2x + 4x$
b $5k + 3k$
c $3t + t$
d $4m - 2m$
e $7k - k$
f $5m - m$

2 Simplify each of these expressions.

a $4m + 3m - 2m$
b $7x + 2x - 3x$
c $10k + 2k - 5k$
d $6p - 3p + p$
e $5a - a + 2a$
f $6y - 4y - y$

3 Simplify each of these expressions.

a $4m + 3m + 2$
b $3x + 2x + 7$
c $7k + 4 + 2k$
d $5b + 2 - 3b$
e $6z - 4 + 2z$
f $10t - 4 - 5t$

4 Simplify each of these expressions.

a $3m + 2m + 2k + 5k$
b $3x + 2x + 7y - 3y$
c $5n + 8m - n - 7m$
d $4t + 9s - 3t - 6s$
e $3m + 2t - 3m - 2t$
f $8w + 3v + 3w + v$

5 Simplify each of these expressions.

a $5ab + 6ab$
b $6w^2 + 3w^2$
c $15xy - 3xy$
d $-3mn + 5mn$
e $-4k - 5k$
f $-y + 10y$

6 Simplify each of these expressions.

a $6m + 4m + 3g + 2g$
b $b + 4b + 3c - 2c$
c $f + 6f + 2g - g$
d $7d^2 + 6de - 2d^2 + 4de$

7 Simplify each of these expressions.

a $2a + 3a + 4b + 2b + 6c + 3c$
b $3m + 2n + 3k + 2m + 5n + 6k$
c $3x + 7y - 2x + 4z - 2y + 3z$
d $15d - 4e + 6e -5d + 8f - 3f$
e $-4u + 6w + v + 10u - 7w + 2v$
f $4r - 3t + 6u +5t - 7r - 8u$
g $-3p - 2q - 3r + 3p + 3r + 7q$
h $3f - 4g + 7f - 7h - 7h + 4g - 10f$
i $2a + a^2 + 3a + 4a^2$
j $4w^2 + 3vw - 2w^2 + 3vw$

Substitution

In sport a substitution is made by **replacing** one player with another.
In mathematics substitution is when you replace a letter, or variable, with a numerical value.
Substituting a different value for a variable will give a different result to the expression.
You need to know how to replace a variable with either a positive or a negative number.

How does that work?

What is the value of $3x + 4$ when $x = 5$?
Replace x with 5.
Remember to put back '×' sign. $= 3 \times 5 + 4$
$= 15 + 4 = 19$

What is the value of $3x + 4$ when x = –2?
Replace x with –2.
Remember to put back '×' sign. $= 3 \times (-2) + 4$
$= -6 + 4 = -2$

What is the value of $3a + 2b$ when a = 6 and b = 4?
Replace each letter with its value.
Remember to put back '×' sign. $= 3 \times 6 + 2 \times 4$
Use **BODMAS** for correct order. $= 18 + 8 = 26$

What is the value of $ab - b^2$ when $a = 7$ and $b = 3$?
Replace each letter with its value.
Remember to put back the '×' sign. $= 7 \times 3 - 3^2$
Use **BODMAS** for correct order. $= 21 - 9 = 12$

What is the value of $\frac{u}{v} + 4v$ when $u = 15$ and $v = 3$?
Replace each letter with its value.
Remember to put back the 'x' sign. $= \frac{15}{3} + 4 \times 3$
Use **BODMAS** for correct order. $= 5 + 12$ $= 17$

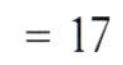

DON'T FORGET

Remember to use 'BODMAS' (Brackets Of Divide, Multipy, Add and Subtract) when doing mixed calculations.

Classroom challenge

1 Calculate the value of each expression, for each given substitution.

a	$x + 4$	when	$x = 3$	$x = 7$	$x = -1$
b	$2x + 3$	when	$x = 1$	$x = 10$	$x = -1$
c	$5 + 7x$	when	$x = 2$	$x = -1$	$x = 4$
d	$3a + 12$	when	$a = 5$	$a = 3$	$a = -4$
e	$15 - 3p$	when	$p = 4$	$p = 5$	$p = -2$
f	$100 - 5k$	when	$k = 10$	$k = 20$	$k = -5$
g	$m^2 + 3m$	when	$m = 3$	$m = 5$	$m = -2$
h	$5w - w^2$	when	$w = 4$	$w = 5$	$w = 7$

2 Given that $a = 4$ and $b = 7$, calculate the value of each expression.

a $2a$
b $3b$
c $2a + 3b$
d $3b - 2a$
e $6a - 3b$
f a^2
g b^2
h $a^2 + b^2$
i $b^2 - a^2$

3 Given that $p = 5$ and $q = -3$, calculate the value of each expression.

a $4p$
b $4q$
c $4p + q$
d $4p + 4q$
e $3p - q$
f $6p - 2q$
g p^2
h $2p^2$
i q^2
j $p^2 + q^2$
k $2p^2 + q^2$
l $5q^2 - 9p$

4 Given that $u = 3$, $v = 5$ and $w = -2$, calculate the value of each expression.

a $u + v$
b uv
c vw
d $2u + v$
e $3uv + w$
f uvw
g uw
h $uw + v$
i $uw - v$
j $uv + vw$
k $uw - 2uv$
l $vw + 4uv$

5 Given that $p = 4$ and $q = 5$, calculate the value of each expression.

a p^2
b q^2
c $p^2 + q^2$
d $p + q^2$
e $p^2 + q$
f pq^2
g p^2q
h $q^2 - p^2$
i $3p^2 + 2q^2$

6 Given that $m = 6$ and $n = -2$, calculate the value of each expression.

a m^2
b n^2
c $m^2 + n^2$
d $m^2 - n^2$
e m^2n^2
f $(mn)^2$

7 Given that $u = 12$ and $v = 3$, calculate the value of each expression.

a $\frac{u}{v}$
b $\frac{u}{v} + 5$
c $\frac{2u}{v}$
d $\frac{u}{2v}$
e $\frac{uv}{9}$
f $\frac{9u}{v^2}$

Substituting in a formula

I can create and evaluate a simple formula representing information contained in a diagram, problem or statement. MTH 3-15b

What's coming up?

This Outcome and Experience will give you the opportunity to:

- create a simple linear formula representing information contained in a diagram, problem or statement
- evaluate a simple formula, for example, $C = 0{\cdot}05m + 75$.

What you already know

You have already learned how to:

- ✔ apply the correct order of operations in number calculations when solving multi-step problems
- ✔ identify familiar contexts in which negative numbers are used
- ✔ extend a given pattern and describe the rule
- ✔ express sequence rules in algebraic notation, for example the cost of hiring a car is £75 plus a charge of £0·05 per mile, '*m*', driven; $C = 0{\cdot}05m + 75$.

Evaluating a formula

You will have seen many different formulae used in a variety of situations.

Quite often, a formula will connect two or more quantities, or allow you to calculate a quantity if you are given another one.

How does that work?

EXAMPLE

The formula for converting inches (i) to centimetres (c) is given by the formula:

$$c = 2{\cdot}54i$$

Convert 10 inches to centimetres.

Step 1 Substitute in formula: $c = 2{\cdot}54 \times 10$

Step 2 Calculate: $= 25{\cdot}4$

So 10 inches = 25·4 centimetres

EXAMPLE

The formula for the area, A, of a rectangle with length, l, and breadth, b, is:

$$A = lb$$

Calculate the area of a rectangle that has a length of 15 cm and a breadth of 7 cm.

Step 1 Substitute in formula: $A = lb$

$= 15 \times 7$

Step 2 Calculate: $= 105\text{ cm}^2$

So area of rectangle $= 105\text{ cm}^2$

EXAMPLE

To convert from degrees Celsius (°C) to degrees Fahrenheit (°F), use the formula

$$F = \frac{9C}{5} + 32$$

Convert 25°C to Fahrenheit.

Step 1 Substitute in formula: $F = \frac{9 \times 25}{5} + 32$

Step 2 Calculate: $= 45 + 32$

$= 77$

So 25°C = 77°F

Classroom challenge

1 Use the formula $A = lb$ to calculate the area of a rectangle which has length 15 cm and breadth 8 cm.

2 Use the formula $A = lb$ to calculate the area of a rectangle with length 4·5 m and breadth 3·2 m.

3 Use the formula $c = 2{\cdot}54i$ to convert 12 inches to centimetres.

4 Use the formula $c = 2{\cdot}54i$ to convert 39 inches to centimetres.

5 Use the formula $F = \frac{9C}{5} + 32$ to convert:

a 30°C to Fahrenheit

b 100°C to Fahrenheit

c 0°C to Fahrenheit

d −40°C to Fahrenheit.

6 The formula to convert degrees Fahrenheit (F) to degrees Celsius (C) is

$$C = \frac{5}{9}(F - 32)$$

Use the formula to convert:

a 104°F to Celsius

b 212°F to Celsius

c 32°F to Celsius

d −40°F to Celsius.

7 Pauline is a plumber. When she is called out to a job she charges a call-out fee of £50 plus £20 per hour. The formula she uses to calculate the bill is therefore:

$C = 20h + 50$

where C is the charge and h is the number of hours spent on the job.
How much will Pauline charge if the job takes:

a 2 hours
b 5 hours
c 3 hours
d $4\frac{1}{2}$ hours?

8 A car hire company uses the formula $C = 0{\cdot}5m + 30$
to calculate the total hire charge for a car, where C is the charge in pounds and m is the mileage covered by the hire car.
What would be the cost to hire a car if the distance travelled is:

a 100 miles
b 250 miles
c 75 miles
d 125 miles?

9 A formula used in physics is $F = ma$.
Find F when:

a $m = 8$ and $a = 10$
b $m = 15$ and $a = 12$
c $m = 6{\cdot}5$ and $a = 5$
d $m = 2{\cdot}5$ and $a = 2{\cdot}5$.

10 The volume of a cuboid is found by using the formula $V = lwh$
where V is the volume, l is the length, w is the width and h is the height of the cuboid.
Calculate V when:

a $l = 10$ cm, $w = 6$ cm, $h = 5$ cm
b $l = 3$ m, $w = 1$ m, $h = 0{\cdot}5$ m
c $l = 5{\cdot}5$ mm, $w = 6$ mm, $h = 12$ mm
d $l = 25$ mm, $w = 15$ mm, $h = 20$ mm.

11 A formula, developed by Albert Einstein is $E = mc^2$.
Use this formula to calculate E when $m = 20$ and $c = 300\,000$.

12 The formula to find the distance (D) travelled, in feet, by a falling body is given by $D = 16t^2$
Where t is the time in seconds.
Find the distance travelled by a falling body after:

a 5 seconds
b 20 seconds
c 8 seconds
d 6·5 seconds.

13 The surface area (S) of a cube is found by using the formula $S = 6s^2$ where s is the length of the side of the cube.
Calculate the surface area of these cubes.

a $s = 5$ cm
b $s = 8$ m
c $s = 3{\cdot}5$ cm
d $s = 4{\cdot}7$ m

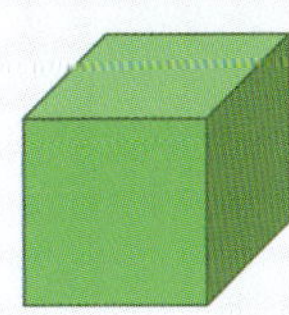

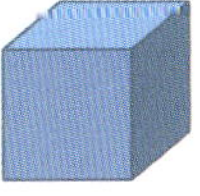

14 A formula used to calculate the distance travelled by a car is $s = ut + \frac{1}{2}at^2$ where:

s is the distance travelled in metres
u is the initial velocity in m/s
a is the acceleration in m/s^2
t is the time in seconds.

Calculate the distance travelled by a car that:

- accelerates at $4\,m/s^2$
- for 6 seconds
- from an initial velocity of $10\,m/s$.

15 Steve dropped a small stone from the top of the Leaning Tower of Pisa in an experiment to calculate the height of the tower.
His friend, Sue, took the following measurements:

u initial velocity = 0 (since it only falls after it was dropped)
a is $10\,m/s^2$ – the acceleration due to gravity
t 3·4 seconds.

Use the formula in question 14 to calculate the height of the Leaning Tower of Pisa, in metres.

Creating a formula: simple linear patterns

Sometimes it is easier to see how a rule is formed if there is a diagram.

The 'Corner Coffee Shop' has triangular tables for customers to sit round.

1 table 2 tables 3 tables

3 customers 6 customers 9 customers

Tables (t)	1	2	3	4	5		10
Customers (c)	3	6	9				

3 3 3 3

For every extra table the number of customers goes up by 3.

In words, the rule is number of customers = 3 times number of tables

Using letters, the rule is $c = 3 \times t$

$c = 3t$

We can use this rule to calculate how many customers would fit around 10 tables.

$c = 3t$

$= 3 \times 10$

$= 30$ customers

Classroom challenge

1 Louise is making squares using lollipop sticks.

1 square
4 lollipop sticks

2 squares
8 lollipop sticks

3 squares
12 lollipop sticks

a Copy and complete this table for the diagrams above.

Squares (s)	1	2	3	4	5
Lollipop sticks (l)	4	8	12		

b Copy and complete: For every extra square the number of lollipop sticks increases by ....

c Copy and complete this rule for finding the number of lollipop sticks.

Number of lollipop sticks = ... times number of squares

$l = \ldots \times s$

$l =$

d How many lollipop sticks will Louise need to make 8 squares?

2 Beatrix is making hexagon shapes using lollipop sticks.

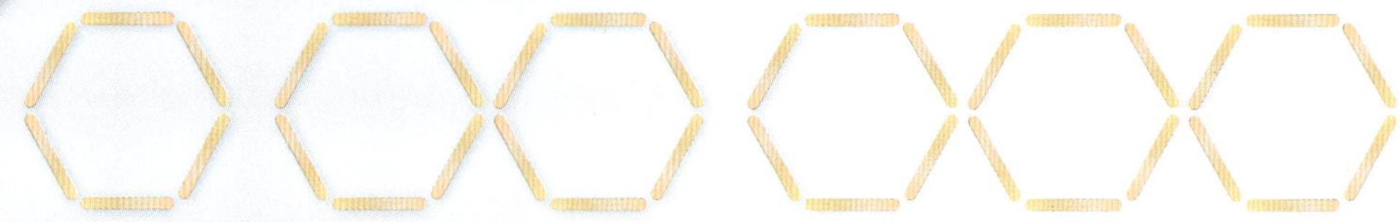

a Copy and complete this table.

Hexagons (h)	1	2	3	4	5
Lollipop sticks (l)	6	12	18		

b Copy and complete: For every extra hexagon the number of lollipop sticks increases by .....

c Copy and complete this rule for finding the number of lollipop sticks.

Number of lollipop sticks = ... times number of hexagons

l = ... × h

l =

d How many lollipop sticks will Louise need to make 7 hexagons?

e How many lollipop sticks will Louise need to make 15 hexagons?

3 Look at his pattern of octopuses and their tentacles.

1 octopus

8 tentacles

2 octopuses

16 tentacles

3 octopuses

24 tentacles

a Copy and complete this table.

Octopuses (p)	1	2	3	4	5
Tentacles (t)	8	16	24		

b Copy and complete: For every extra octopus the number of tentacles increases by ....

c Copy and complete this rule for finding the number of tentacles.

Number of tentacles = ... times the number of octopuses

t = ... × p

t =

d How many tentacles would there be if there were 20 octopuses?

4

1 textbook
£20

2 textbooks
£40

3 textbooks
£60

a Copy and complete this table.

Textbook (t)	1	2	3	4	5
Cost (c)	20	40	60		

b Copy and complete: For every extra textbook the cost increases by ... pounds.

c Copy and complete this rule for finding the cost of a number of textbooks.

Cost of textbooks = ... times the number of books

c = ... × t

c =

d What would be the cost of 30 textbooks?

5 For each of these tables, determine a rule or formula connecting the two letters.

a

Vases (v)	1	2	3	4	5
Flowers (f)	7	14	21		

b

Cans (c)	1	2	3	4	5
Millilitres (m)	110	220	330		

c

Miles walked (m)	1	2	3	4	5
Time taken in minutes (t)	15	30	45		

d

Tables (t)	1	2	3	4	5
People (p)	9	18	27		

e

No of hours (h)	1	2	3	4	5
No. of minutes (m)	60	120	180		

More linear patterns

Look at this pattern.

How does that work?

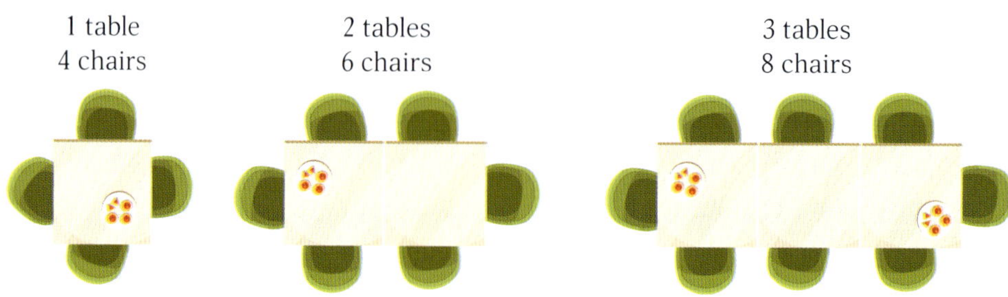

Construct a table like before.

No of tables (t)	1	2	3	4	5
No. of chairs (c)	4	6	8	10	12

+2 +2 +2

For every extra table, the number of chairs increases by 2.
From the last exercise, we would try multiplying by 2.

Check:
$2 \times 1 = 2$ actual number of chairs = 4
$2 \times 2 = 4$ actual number of chairs = 6
$2 \times 3 = 6$ actual number of chairs = 8
$2 \times 4 = 8$ actual number of chairs = 10
$2 \times 5 = 10$ actual number of chairs = 12

There looks to be another step to get from number of tables to number of chairs.
Look at the 3rd row above.

$2 \times 3 = 6$ actual number of chairs = 8
$2 \times 3 = 6$ how do we get to 8? We **add 2**.
$2 \times 3 = 6 + 2 = 8$

The rule is Number of chairs = 2 times number of tables add 2

$$c = 2 \times t + 2$$
$$c = 2t + 2$$

Check for 4 tables

$$c = 2 \times 4 + 2$$
$$= 8 + 2$$
$$= 10 \text{ chairs}$$ correct!

EXAMPLE

Look at the table below. Construct a rule connecting c and h.

No of hours (h)	1	2	3	4	5
Cost in £ (c)	2	5	8	11	14

+3 +3 +3 +3

For each extra hour, the cost increases by 3 => × 3

Check: $3 \times 1 = 3$ actual cost is 2 need to **subtract 1**

Check for one of the other pairs of values in the table.

$3 \times 5 = 15$ $15 - 1 = 14$ correct!

Rule is $c = 3h - 1$

What would be the cost for a time of 8 hours?

Rule is $c = 3h - 1$

$h = 8$ $c = 3 \times 8 - 1$

$= 23$

Classroom challenge

1 A pattern is made using lollipop sticks, as shown below.

a Draw the next set of lollipop stick pattern with 4 squares.

b Copy and complete this table.

No of squares (s)	1	2	3	4	5	6
No of lollipop sticks (l)	4	7	10			

c For every extra square, what is the increase in the number of lollipop sticks?

d Copy and complete this formula for calculating the number of matches required if you know the number of squares.

Formula:

$l = \ldots \times s + \ldots$

e Use your formula to calculate how many lollipop sticks are needed to make 10 squares.

f How many lollipop sticks would be needed to make 15 squares?

STRETCH YOURSELF

g How many squares could be made if there were 61 lollipop sticks?

2 Here are some hexagon patterns made using sticks.

1 hexagon 6 sticks | 2 hexagons 11 sticks | 3 hexagons 16 sticks

a Draw the next pattern in the sequence.

b Copy and complete this table.

No of hexagons (h)	1	2	3	4	5	6
No of sticks (s)	6	11	16			

c For every extra hexagon the number of sticks increases by … .

d Copy and complete this formula for calculating the number of sticks needed, given the number of hexagons.

$s = \ldots \times h + \ldots$

$s = \ldots\ h + \ldots$

e How many sticks would be needed for a pattern with 10 hexagons?

f How many sticks would be needed for a pattern with 25 hexagons?

STRETCH YOURSELF

g How many hexagons could be made with 91 sticks?

3 Mark makes these patterns using paper straws.

a Draw the next 2 patterns in the sequence.

b Copy and complete this table.

No of triangles (t)	1	2	3	4	5	6
No of straws (s)	3	5	7			

c For every extra triangle the straws increase by .... .
d Copy and complete the formula $s = \dots t + \dots$
e How many straws would be needed for a pattern with 8 triangles?
f How many straws would be needed for a pattern with 12 triangles?

STRETCH YOURSELF

g How many triangles could Mark make if he had 101 straws?

4 Sohail sees an advert which shows the cost of hiring a wallpaper stripper.

No of days hired (d)	1	2	3	4	5	6
Cost in £ (c)	9	19	29			

a Copy and complete the table for hiring the wallpaper stripper for 4, 5 and 6 days.
b How much extra does the hire cost for each extra day?
c Copy and complete this formula for calculating the cost of hire, given the number of day.

$c = \dots \times d - \dots$

$c = \dots d - \dots$

d Use your formula to calculate how much it will cost to hire the wallpaper stripper for:
i 10 days ii 14 days iii 21 days.

STRETCH YOURSELF

e Sohail has £79. For how many days could he hire the wallpaper stripper?

5 The number of litres of water which need to be added to cupfuls of grass seed is shown in the table.

Cupfuls of grass seed (g)	1	2	3	4	5	6
Litres of water (w)	0·4	0·9	1·4	1·9		

a For every extra cupful of grass seed, the number of litres of water increases by ....
b Copy and complete this formula $w = \dots g - \dots$
c How many litres of water would be needed for 18 cups of grass seed?

STRETCH YOURSELF

d A watering can holds 10 litres of water. Would this be enough for 5 cupfuls of grass seed?

6 Here is a pattern made from bricks.

a Draw the 4th pattern.

b Copy and complete this table for the pattern.

Pattern (p)	1	2	3	4	5	6
Number of bricks (b)	5	9	13			

c Using b for bricks and p for pattern, write a formula for calculating the number of bricks required for a given pattern: $b = \ldots\ p \ldots$

d How many bricks would be required for the 9th pattern?

e Which pattern will have 49 bricks?

7 For each of the tables below, write a formula connecting the second letter to the first.

a

Distance (d)	1	2	3	4
Litres of fuel (f)	3	5	7	9

$f = \ldots\ d + \ldots$

b

No of packets (p)	1	2	3	4
No of teabags (t)	40	60	80	100

c

No of plants (p)	1	2	3	4
No of berries (b)	12	17	22	27

d

No of days hired (d)	1	2	3	4
Cost in £ (c)	12	27	42	57

e

No of boxes (b)	1	2	3	4
No of chocolates (c)	8	18	28	38

Creating a formula from a diagram or problem

Sometimes you may be asked to write down a rule or formula from a diagram to help you to solve a problem.

How does that work?

Write down a formula for:

a the perimeter of this rectangle

b the area of this rectangle.

m

n

SOLUTION

Perimeter (P) $= m + n + m + n$

$P = 2m + 2n$ or $2(m + n)$

Area (A) $= m \times n$

$A = mn$

These formulae can now be used to calculate the perimeter or area when given values for m and n.
Find the perimeter and area of the rectangle when $m = 5\,\text{cm}$ and $n = 3\,\text{cm}$.

$$P = 2m + 2n \qquad A = mn$$
$$= 2 \times 5 + 2 \times 3 \qquad = 5 \times 3$$
$$= 16\,\text{cm} \qquad = 15\,\text{cm}^2$$

How does that work?

EXAMPLE

Jayne, a landscape gardener, builds a path using orange and grey slabs, as shown.
The pattern is a row of grey slabs on each side of the orange slabs.
If o orange slabs were used, how many grey (g) slabs would be needed?

SOLUTION

Look at how the pattern builds up.

1 orange slab 2 grey slabs

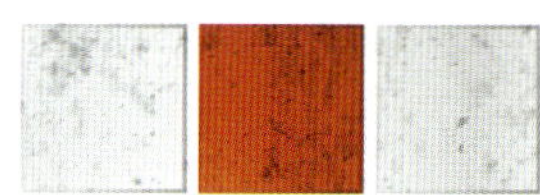

2 orange slabs 4 grey slabs

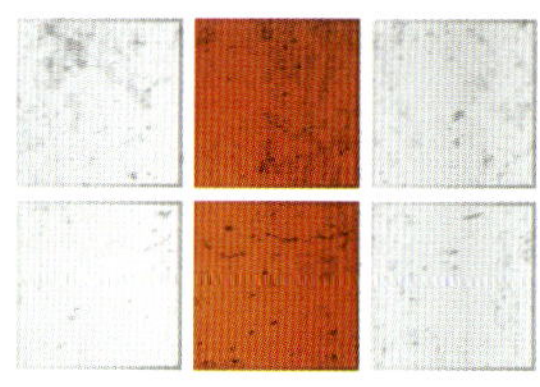

3 orange slabs 6 grey slabs

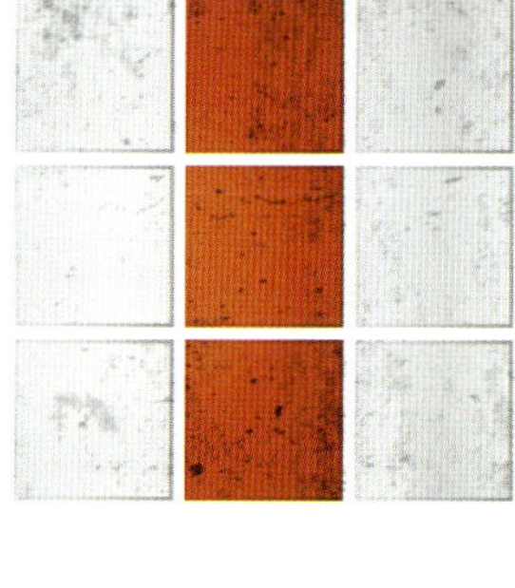

4 orange slabs 8 grey slabs

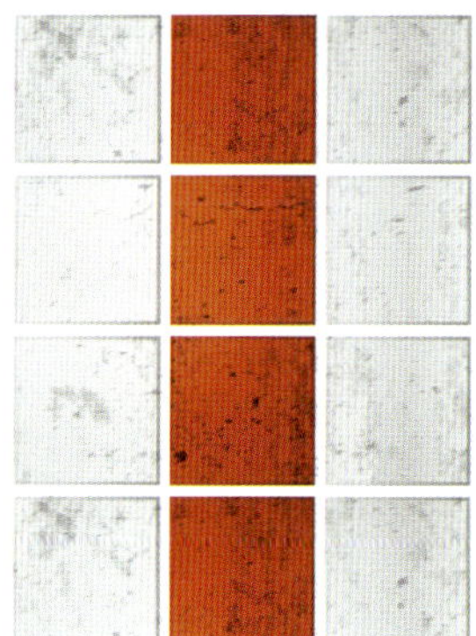

For **each** orange slab there are 2 grey slabs.
That is, number of grey slabs (g) = 2 × orange slabs (o)
The formula is $g = 2o$

If Jayne used 15 orange slabs, how many grey slabs would she need?

SOLUTION

Formula $g = 2o$
Substitute $= 2 \times 15$
Calculate $= 30$ Jayne would need 30 grey slabs.

EXAMPLE

3 consecutive numbers are added together.
If the first number is n, write down expressions for the next two.
Write down an expression for the total of the 3 numbers.

Consecutive numbers are numbers that follow on by increasing by 1 each time. For example, 1, 2, 3 or 400, 401, 402 or 18, 19, 20.

SOLUTION

First number/expression is	n	(told to use n as first number)
Second number/expression is	$n + 1$	(consecutive numbers so +1)
Third number/expression is	$n + 2$	
Total	$3n + 3$	

Classroom challenge

1 Write down a formula for the perimeter of each of these shapes.
Your formula should start $P =$

a

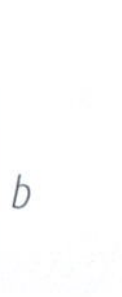

b

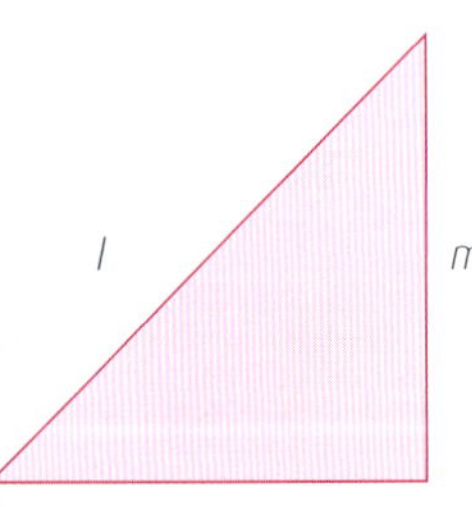

c

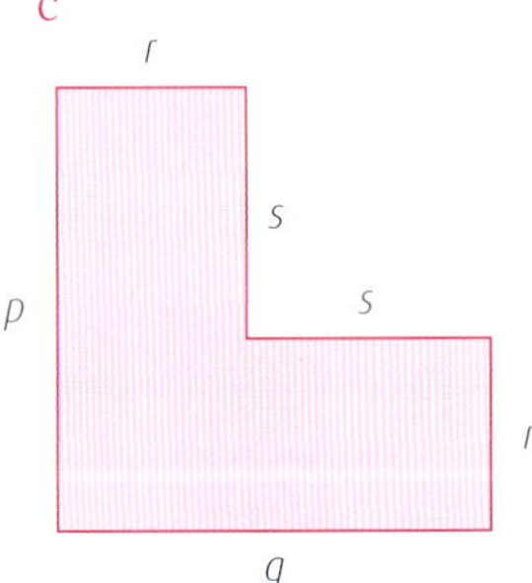

2 Construct a formula for the area of each of these shapes.
Use your formula to calculate the area for the values given.

a

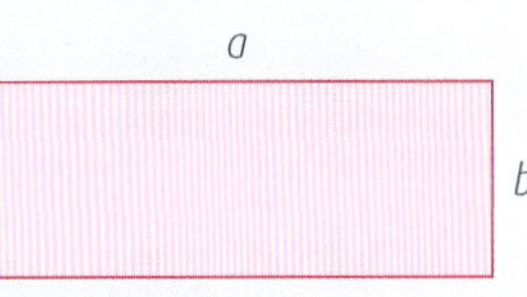

$a = 6\,\text{cm}$, $b = 10\,\text{cm}$

b

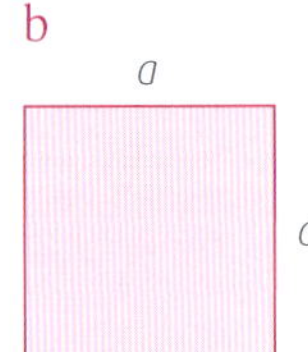

$a = 3\,\text{cm}$

c

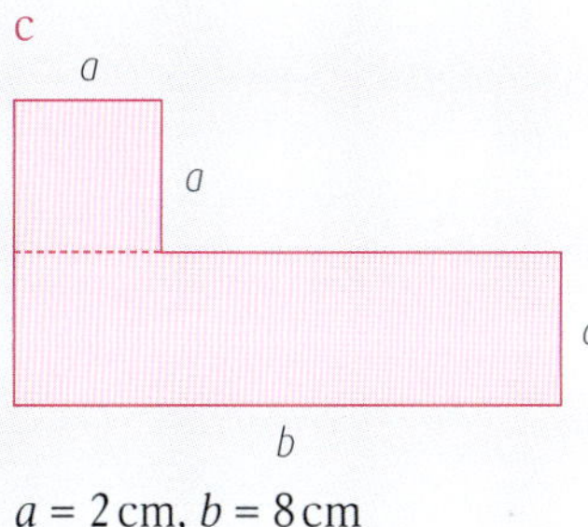

$a = 2\,\text{cm}$, $b = 8\,\text{cm}$

d

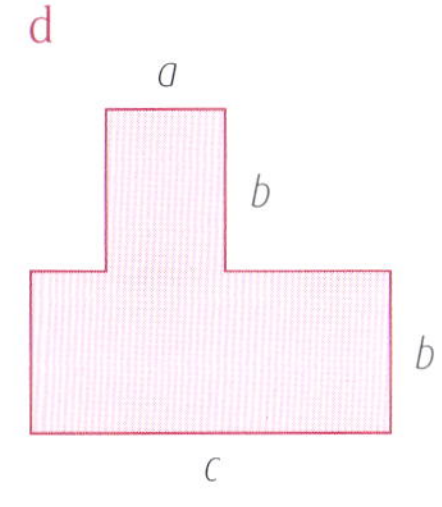

$a = 3\,\text{cm}$, $b = 4\,\text{cm}$, $c = 9\,\text{cm}$

3 Three consecutive numbers are to be added together.
   a If n is the first number, write down an expression for the next two numbers.
   b Show that this **total** is always divisible by 3.

4 The ferryman at Cramond, Edinburgh, can only carry one passenger at a time on his ferry across the river.

1 person
1 crossing

2 people
3 crossings

3 people
5 crossings

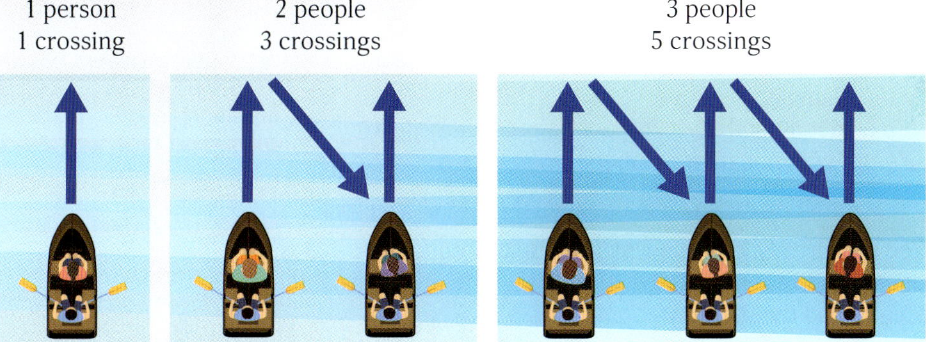

How many crossings would it take to ferry 8 people across the river?

5 Sammy is 5 years older than Miguel.
Together their ages total 27 years.
How old is Miguel?

6 Mirror tiles are surrounded by ceramic tiles, as shown.

1 mirror tile

2 mirror tiles

3 mirror tiles

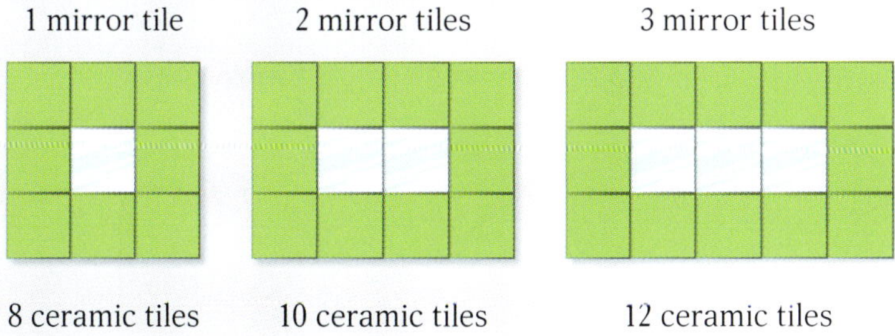

8 ceramic tiles

10 ceramic tiles

12 ceramic tiles

   a Draw the next pattern.
   b Write down a formula for calculating the number of ceramic tiles (c) required if you know the number of mirror tiles (m).

Equations

Having discussed ways to express problems or statements using mathematical language, I can construct, and use appropriate methods to solve, a range of simple equations. MTH 3-15a

What's coming up?

This Outcome and Experience will give you the opportunity to:

- solve linear equations, for example, $ax \pm b = c$, where a, b and c are integers.

What you already know

You have already learned how to:

- ✔ solve simple algebraic equations with one variable, for example, $a - 30 = 40$ and $4b = 20$
- ✔ work with negative numbers.

DON'T FORGET

Reminder
1-step equations

Solve	x	+	3	=	8
Step 1		−	3		− 3
	x			=	**5**

Solve	$3x$	=	12
Step 1	÷ 3		÷ 3
	x	=	**4**

You will already know how to solve '1-step' equations, that is, equations like $x + 4 = 7$ and $3x = 12$ where it only takes one step to find the solution.

In this section, we will look at how to solve '2-step' equations.

2-step equations combine these operations:

EXAMPLE

Solve	$3x$	+	4	=	19
Step 1		−	4		− 4
	$3x$			=	15
Step 2	÷ 3				÷3
	x			=	5

EXAMPLE

Solve	$4g$	−	7	=	13
Step 1		+	7		+ 7
	$4g$			=	20
Step 2	÷ 4				÷ 4
	g			=	5

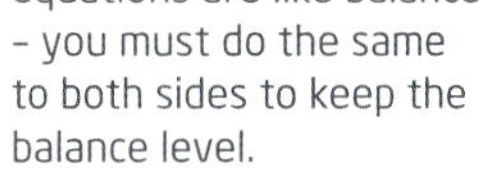

DON'T FORGET

Remember:
equations are like balances – you must do the same to both sides to keep the balance level.

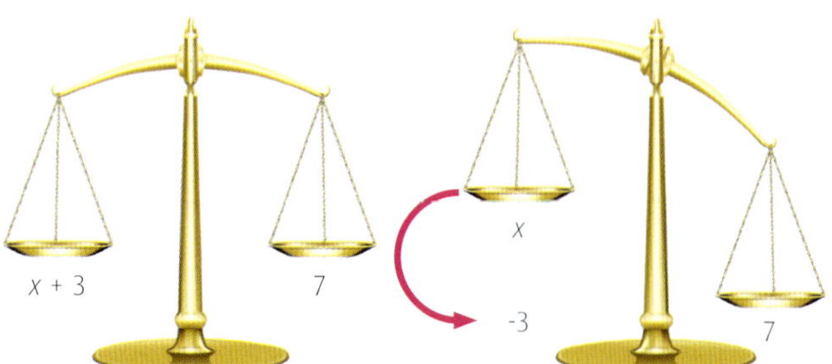

You need to -3 from this side to keep the balance

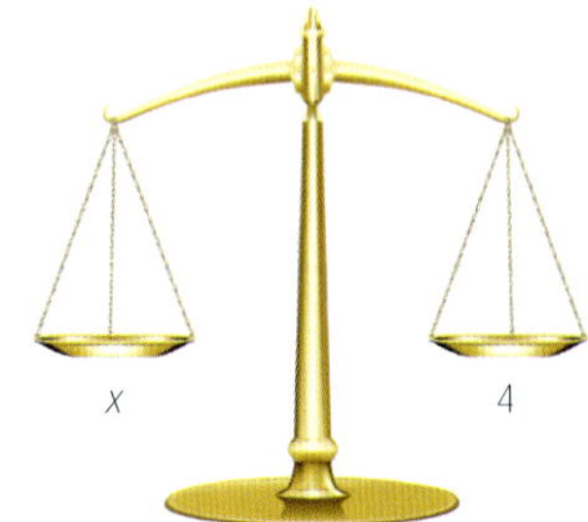

Classroom challenge

Warm up!

Some 1-step equations to get your brain into gear!

1 Solve these equations.

a $x + 4 = 7$
b $x + 3 = 9$
c $x + 7 = 11$
d $x - 4 = 7$
e $x - 3 = 9$
f $x - 7 = 11$
g $a - 2 = 0$
h $b + 3 = 3$
i $c + 4 = 3$
j $d + 7 = 5$
k $k - 2 = -1$
l $m - 4 = -7$

2 Solve these equations.

a $3x = 12$
b $5x = 15$
c $4x = 28$
d $7x = 21$
e $9x - 27$
f $2x = 40$
g $8a = 32$
h $10h = 80$
i $3m = 33$
j $11n = 66$
k $5p - 15$
l $3q = -24$

Warmed up? Let's go!

3 Solve these equations.

a $2x + 1 = 11$
b $3x + 5 = 17$
c $5x + 3 = 23$
d $6a + 4 = 22$
e $7g + 2 = 30$
f $4h + 10 = 18$
g $2x - 1 = 9$
h $3x - 4 = 11$
i $5x - 2 = 23$
j $4a + 3 = 15$
k $6g + 8 = 32$
l $9h + 9 = 9$
m $7t - 3 = 32$
n $6t - 3 = 45$
o $2k - 3 = 11$
p $2x + 3 = 8$
q $2x - 5 = 2$
r $4x + 7 = 21$

4 Solve these equations.

a $4 + 2x = 10$
b $5 + 3x = 17$
c $7 + 4x = 35$
d $10 - 2x = 0$
e $15 - 3x = 12$
f $8 - 2x = 2$

5 Solve these equations.

a $2x + 6 = 2$
b $5x + 13 = 3$
c $3 + 2x = 1$

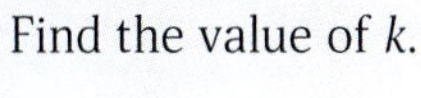

STRETCH YOURSELF

6 Solve these equations.

a $\frac{x}{2} = 6$
b $\frac{x}{2} + 4 = 10$
c $\frac{x}{2} - 7 = 0$
d $\frac{x}{3} + 4 = 9$
e $\frac{x}{3} - 2 = 4$
f $\frac{x}{3} + 8 = 8$

Puzzle

Can you solve this puzzle?

Given $A + A = B$
$B + B = D$
$A + D = E$

Then $A + D + E = k\text{B}$
Find the value of k.

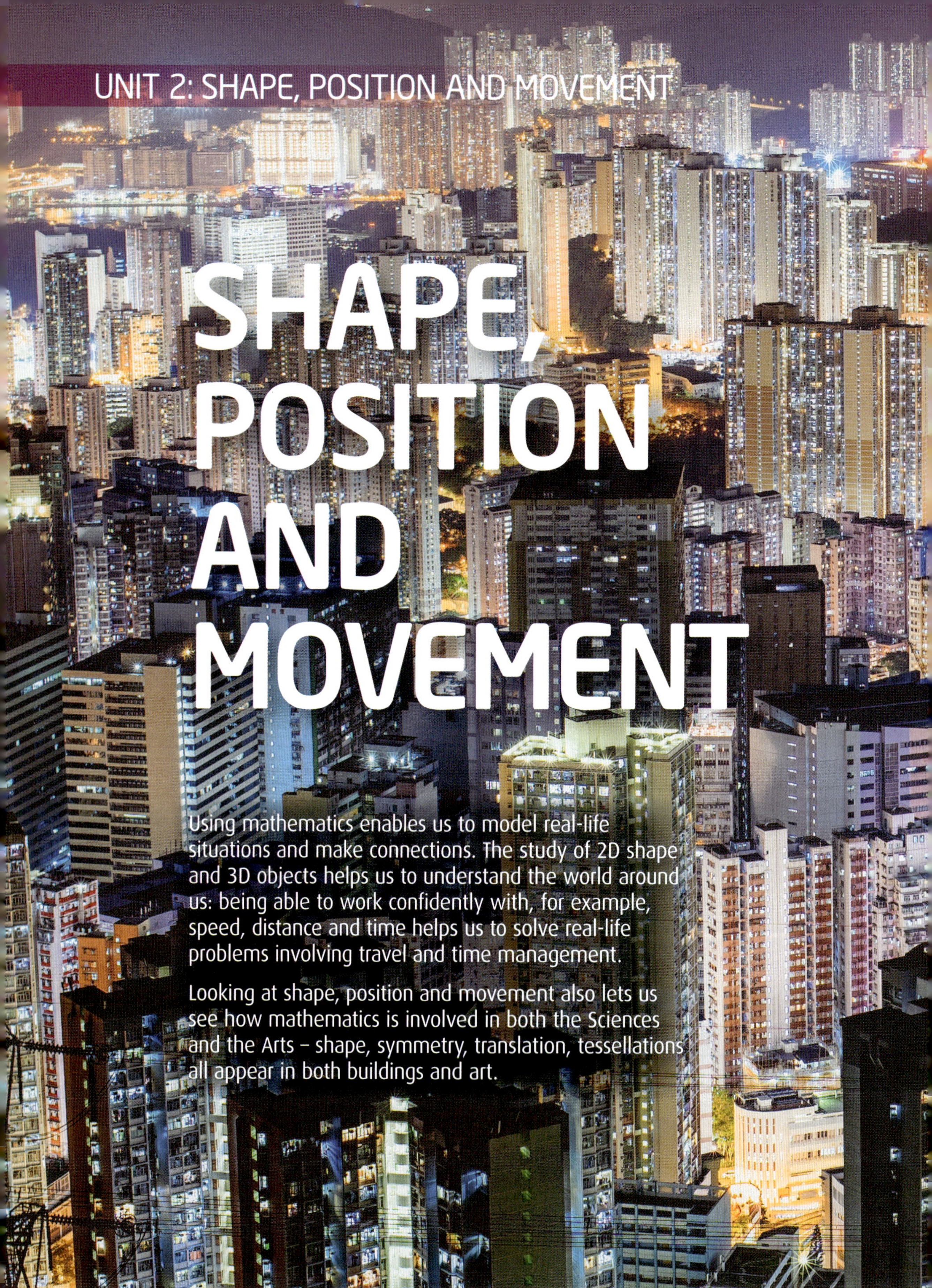

SHAPE, POSITION AND MOVEMENT

Using mathematics enables us to model real-life situations and make connections. The study of 2D shape and 3D objects helps us to understand the world around us: being able to work confidently with, for example, speed, distance and time helps us to solve real-life problems involving travel and time management.

Looking at shape, position and movement also lets us see how mathematics is involved in both the Sciences and the Arts – shape, symmetry, translation, tessellations all appear in both buildings and art.

UNIT 2: SHAPE, POSITION AND MOVEMENT

SHAPE

PROPERTIES OF 2D SHAPES AND 3D OBJECTS

Having investigated a range of methods, I can accurately draw 2D shapes using appropriate mathematical instruments and methods. MTH 3-16a

What's coming up?

This Outcome and Experience will give you the opportunity to:

- demonstrate a variety of methods to accurately draw 2D shapes, including triangles and regular polygons (given the interior angle), using mathematical instruments.

What you already know

You have already learned how to:

- ✓ describe 3D objects and 2D shapes using specific vocabulary including regular, irregular, diagonal, radius, diameter and circumference
- ✓ apply this knowledge to demonstrate understanding of the relationship between 3D objects and their nets
- ✓ identify and describe 3D objects and 2D shapes within the environment and explain why their properties match their function
- ✓ know that the radius is half of the diameter
- ✓ use digital technologies and mathematical instruments to draw 2D shapes and make representations of 3D objects, understanding that not all parts of the 3D object can be seen.

Drawing 2D shapes

Why do I need to learn about shape?

Every day we are surrounded by space and shape.

The world is full of shapes of different sizes – it is part of a universe full of stars and planets of different shapes and sizes.

To understand the world and our environment, we need to have an understanding of shape, position and movement.

Shape also appears in everyday life, for example 'Put a cross in the box' – 'put a tick in the circle' when filling in forms.

In this section, you will practise your 'spacial awareness'.

How does that work?

To draw 2D shapes, you will need to be able to use the following.

A ruler	
A protractor	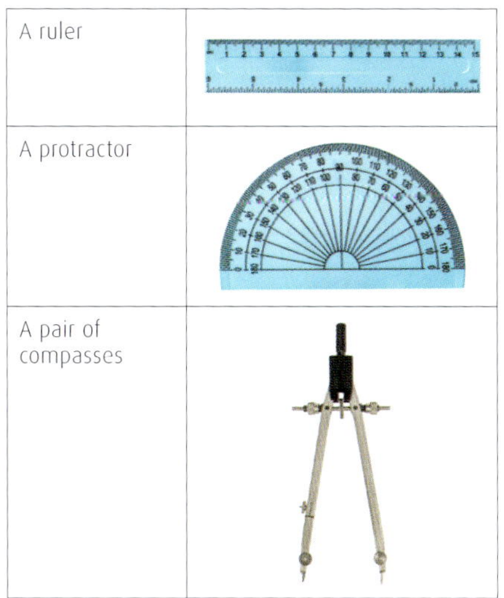
A pair of compasses	

If available, a set square	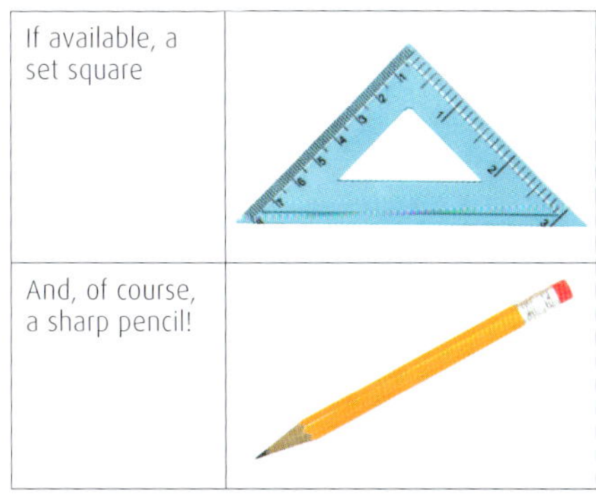
And, of course, a sharp pencil!	

Drawing triangles

Triangles appear in many constructions, mainly because a triangle is a very 'strong' shape.

In this section, you will learn how to draw triangles given different pieces of information.

You should be as accurate as possible:

- when drawing a line, you should draw it to within 2 mm
- when measuring an angle, you should be accurate to within 2°.

How does that work?

INFORMATION 1: 2 SIDES AND 1 ANGLE

Here is a sketch of a triangle ABC. It is not drawn accurately.

Draw this triangle accurately.

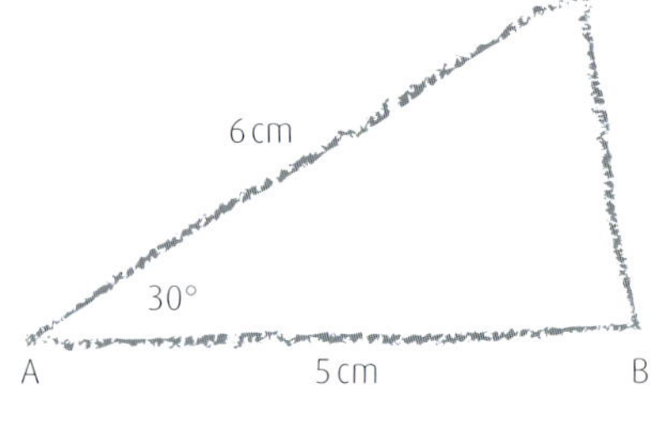

SOLUTION

Step 1 Draw line AB 5 cm long.

A B

Step 2 Measure the angle at A to be 30° and mark it.

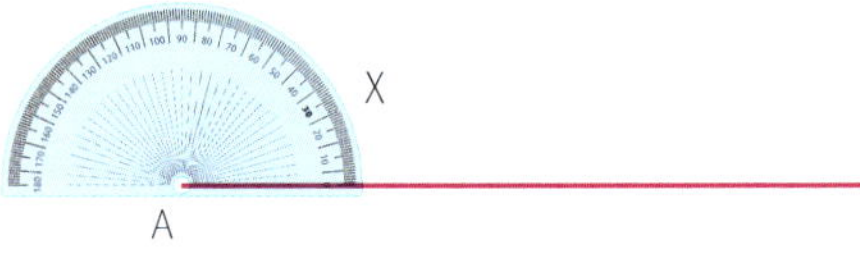

Step 3 Draw line AC 6 cm long.

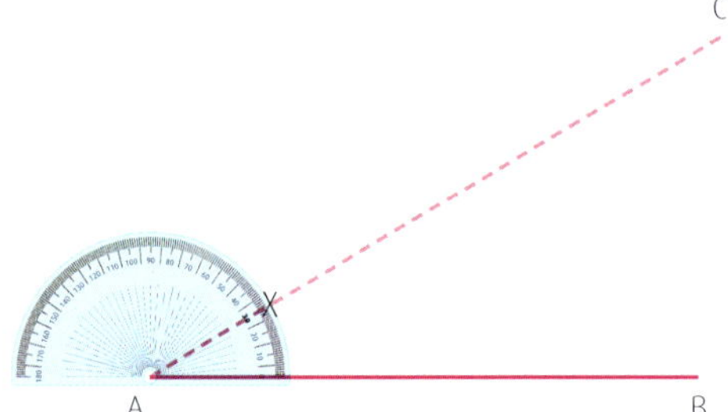

Step 4 Join C to B to complete triangle.

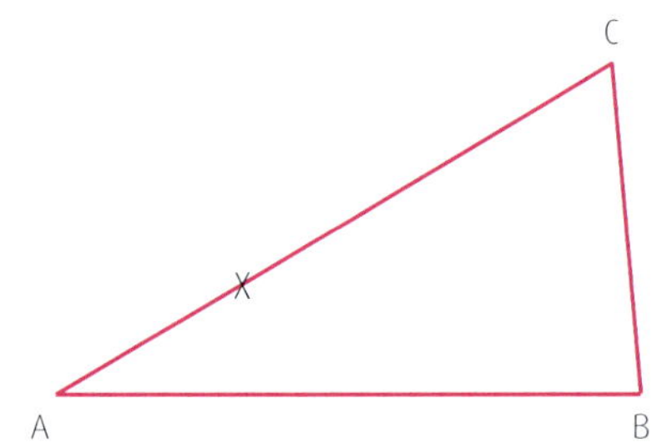

DON'T FORGET

Here you have drawn a Side, an Angle, a Side:
SAS **2 sides** and enclosed **angle**.

INFORMATION 2: 1 SIDE AND 2 ANGLES

EXAMPLE

Here is a sketch of triangle ABC. It is not drawn accurately.
Draw this triangle accurately.

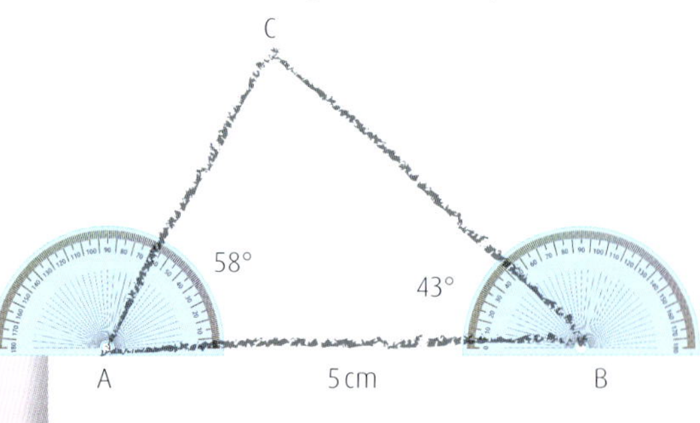

SOLUTION

Step 1 Draw line AB 5 cm long.
Step 2 Draw angle BAC to be 58° (draw feint line).
Step 3 Draw angle ABC to be 43°.
Step 4 Draw line to intersect line in step 2.

DON'T FORGET

Here you have drawn an Angle, Side, Angle:
ASA **2 angles** and 'enclosed' **side**.

INFORMATION 3: 3 SIDES

EXAMPLE

Here is a sketch of triangle ABC. It is not drawn accurately.
Draw this triangle accurately.

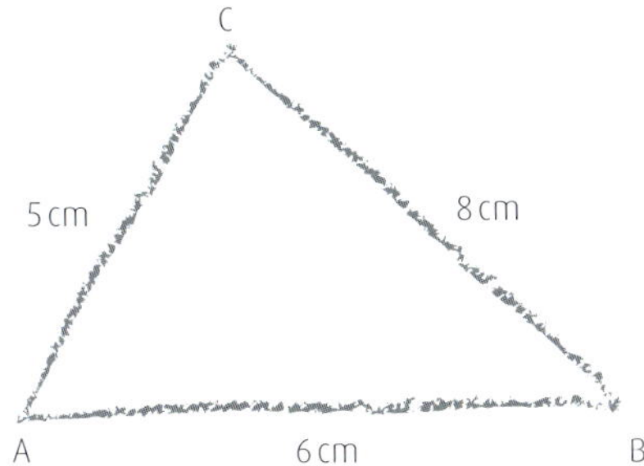

SOLUTION

Step 1 Draw line AB 6 cm long.

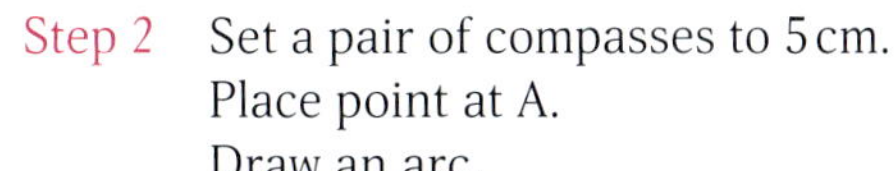

Step 2 Set a pair of compasses to 5 cm.
Place point at A.
Draw an arc.

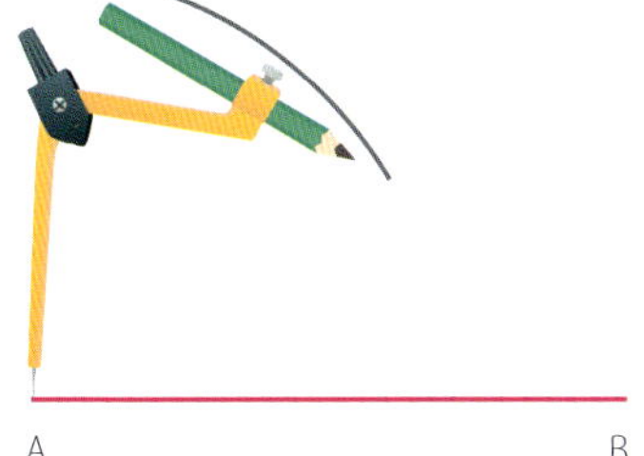

Step 3 Set a pair of compasses to 8 cm.
Place at point B.
Draw an arc to cross first arc.

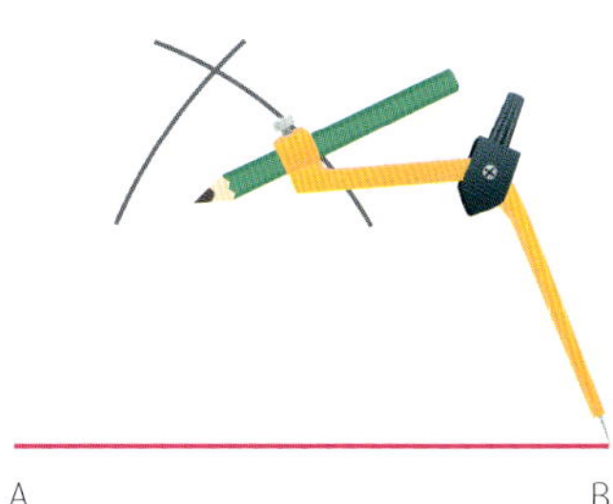

Step 4 Join A to point of intersection.
Join B to point of intersection.
Rub out arcs.

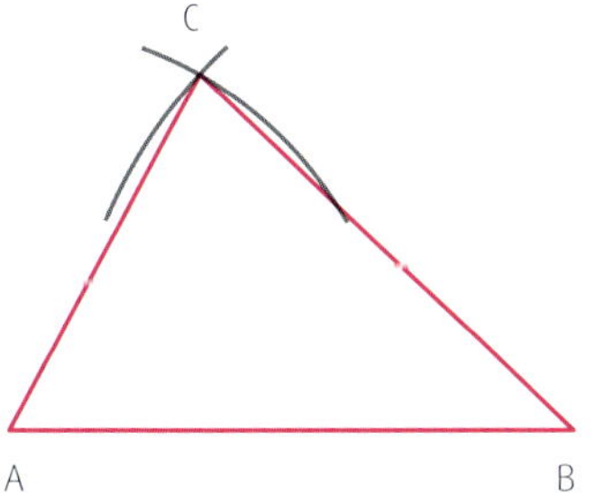

Here you have drawn a
Side, **S**ide, **S**ide:
SSS 3 sides.

Classroom challenge

Here are sketches of triangles. They have not been drawn accurately. Draw these accurately, from the information given.

1 SAS

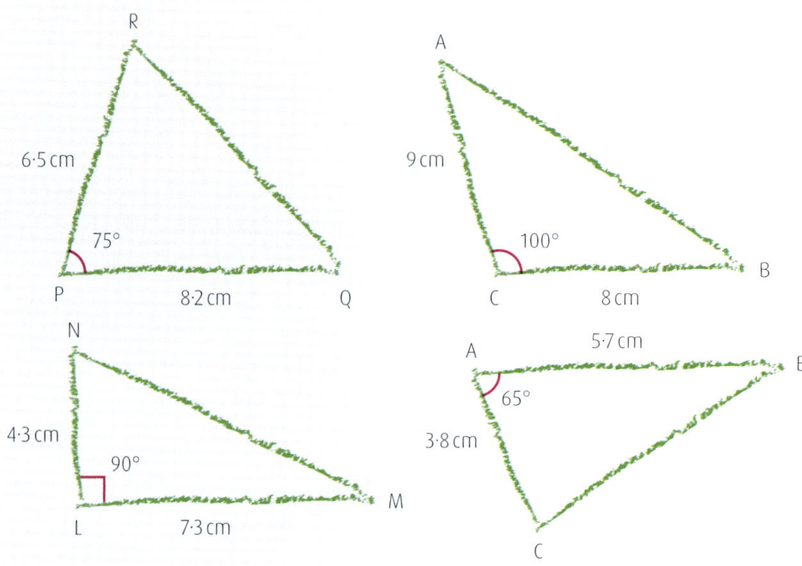

2 ASA

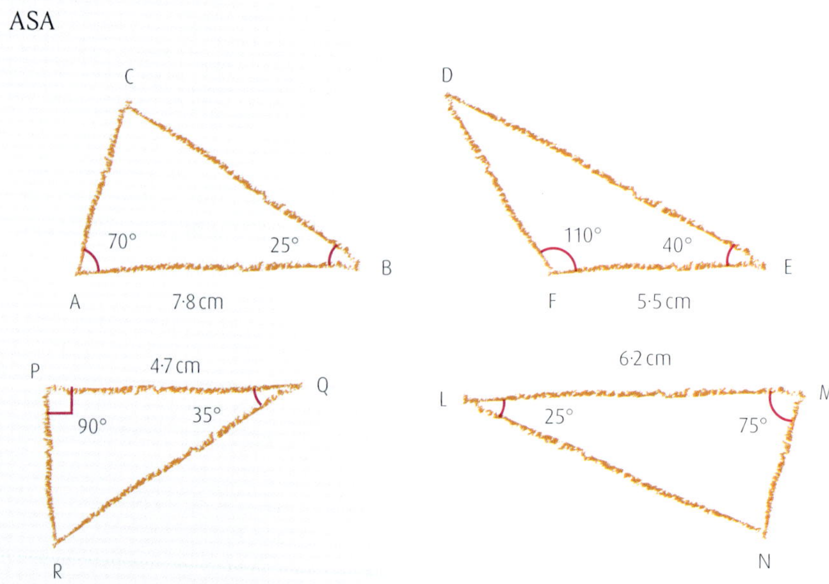

3 SSS

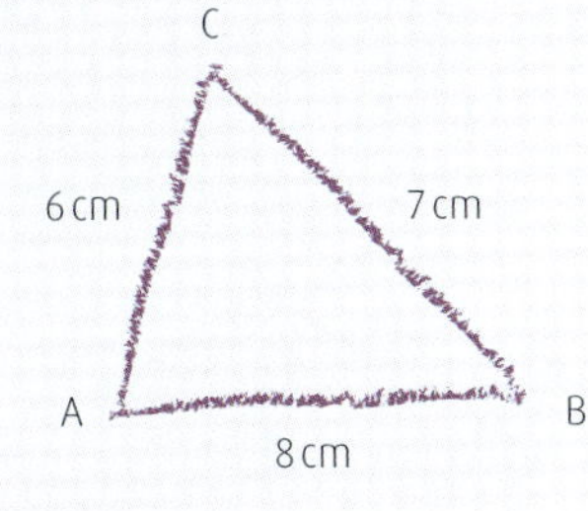

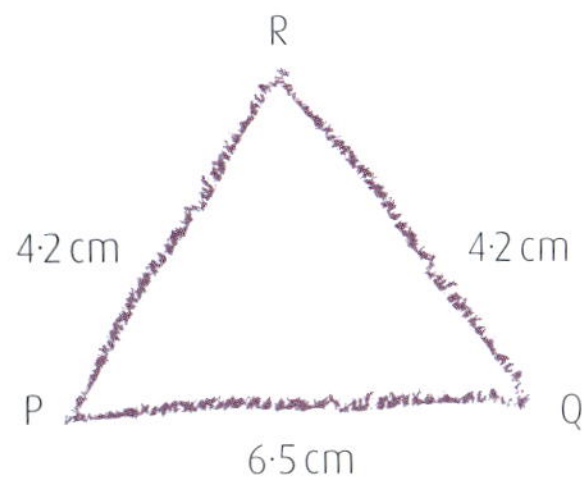

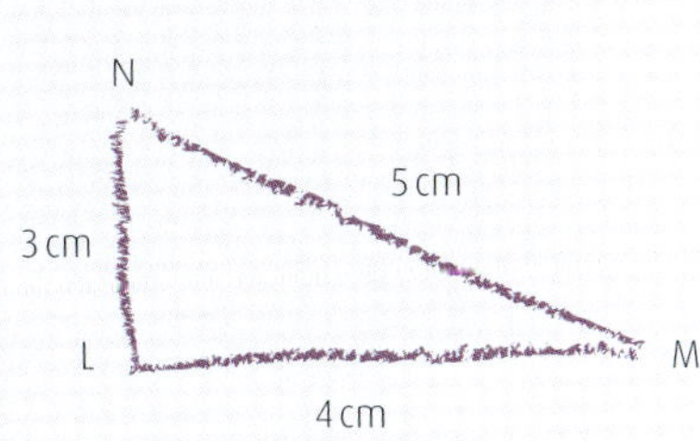

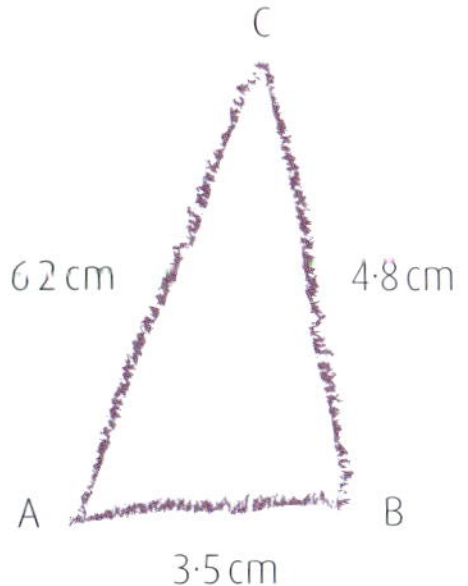

4 Construct each of these triangles from the information given.

a Construct a triangle STU in which ∠T = 60°, ∠U = 70° and TU = 7·5 cm.

b Construct a triangle ABC in which ∠C = 90°, ∠B = 45°, CB = 5 cm.

c Construct a triangle PQR in which AB = 6 cm, AC = 4 cm and ∠A = 35°.

d Construct a triangle LMN in which LM = 53 mm, LN = 62 mm and ∠L = 98°.

e Construct a triangle RST in which RS = 4·7 cm, RT = 6·1 cm and ST = 5·1 cm.

f Construct a triangle TUV in which TU = 60 mm, TV = 80 mm and UV = 100 mm.

g Construct an equilateral triangle ABC in which each side is 5 cm long. Measure the size of each angle; what is the size of each angle?

h Construct a right-angled triangle in which the sides around the right angle are 5 cm and 12 cm. What is the length of the third side of the triangle?

i Construct a triangle PQR in which PQ = PR = 5·6 cm, and QR = 3·3 cm.

STRETCH YOURSELF

5 Fergus thinks he can draw a triangle with sides 4 cm, 5 cm and 10 cm long. He tries but cannot complete the triangle.

a Use the information given to try to draw Fergus's triangle.

b Explain why the triangle cannot be drawn.

Download the Teaching Notes from www.brightredpublishing.co.uk for an extra activity on how to create the Star of David

Constructing other 2D shapes: regular polygons

How does that work?

Construct a regular hexagon

SOLUTION

Step 1 Draw a circle, radius 5 cm.
Step 2 Keep compasses at same setting.
Put point on circumference of circle.
Mark in an arc.
Step 3 Put point of compasses on this arc.
Mark in another arc.
Step 4 Repeat this until you get back to the start.
Step 5 Use a ruler and join up the points in order.
You should now have a regular hexagon.

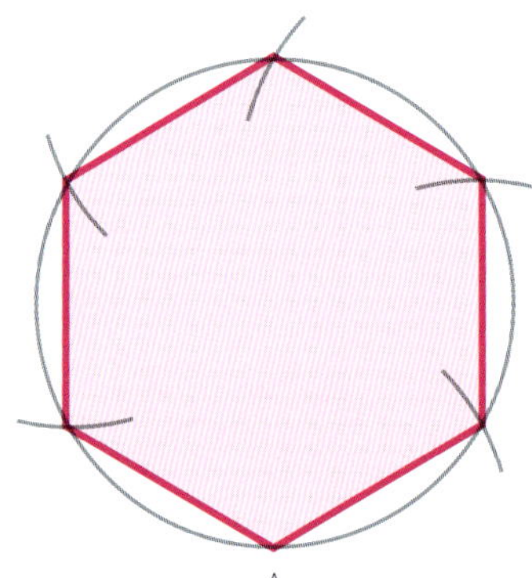

Classroom challenge

1 Complete these steps to draw a regular pentagon.
Step 1 Draw a circle of radius 5 cm.
Step 2 Mark in one radius.
Step 3 Measure an angle of 72° at the centre.
Draw in another radius.
Step 4 Repeat until you get back to the start.
Step 5 Join the points in order.

a How many sides does a pentagon have?
b How many degrees are there in a complete circle?
Do you see that 72° came from 360 ÷ 5?

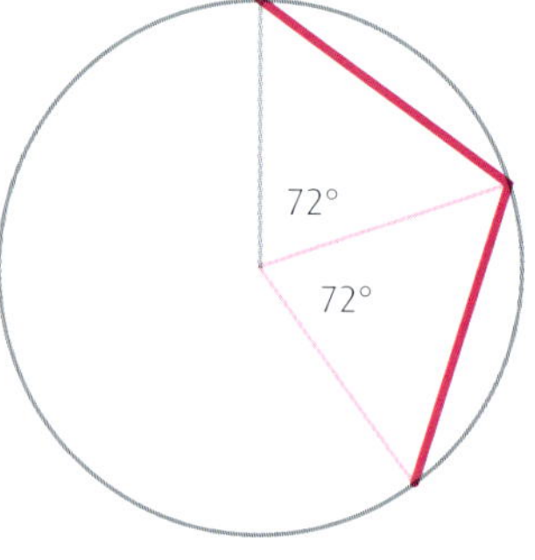

2 For a regular octagon.
There are 8 sides in an octagon:
360 ÷ 8 = 45°
This time the angle at the centre will be 45°.

3 Construct a regular decagon (10-sided shape).

Constructing other 2D shapes

Classroom challenge

Using your mathematical equipment, draw these shapes.

1 a Construct this trapezium.
  b Measure the size of angle Q.
  c Measure the lengths of lines PQ and QR.

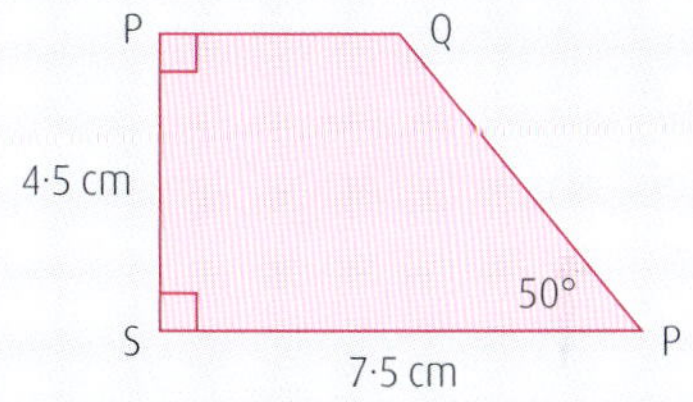

2 Construct this rhombus.

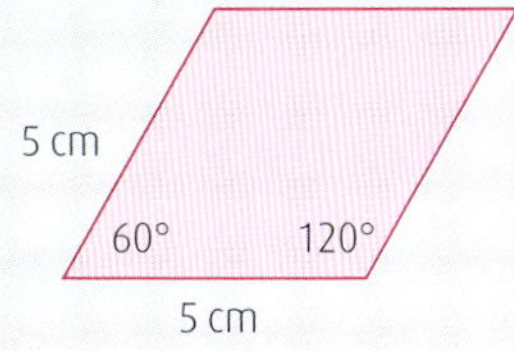

3 Construct this parallelogram.

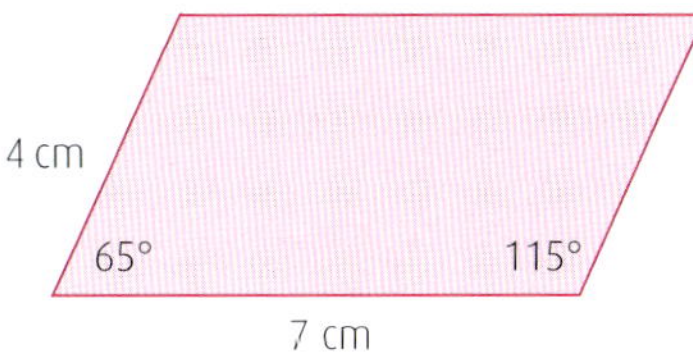

4 Construct this quadrilateral.

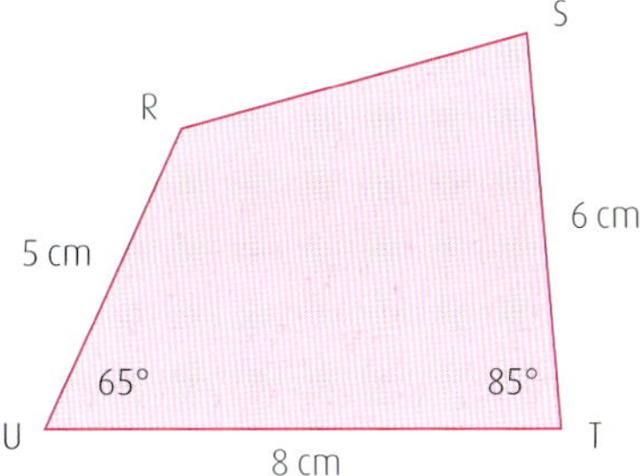

Constructing nets of 3D objects

A 'net' or 'development' is the shape you get when you 'fold out' a 3D object.

How does that work?

If you fold out, or open up a cuboid, you get a net like this one.

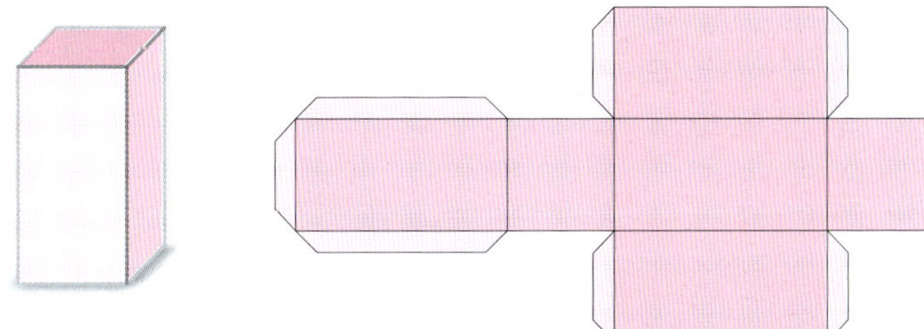

In reverse, if you want to make a 3D object, you should start by constructing the net. The net should have tabs (lighter bits in net above) so that you can glue the sides together.

Classroom challenge

Draw each of these nets accurately, on card.

You can then cut out the net and fold it to make the 3D object.

1 Square-based pyramid

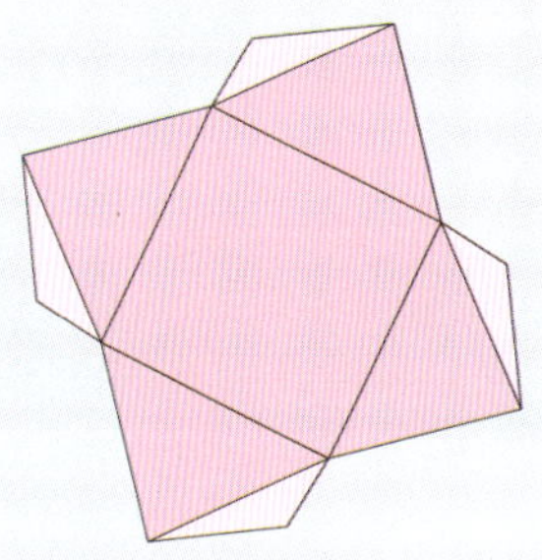

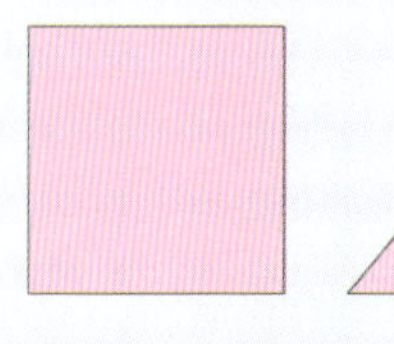

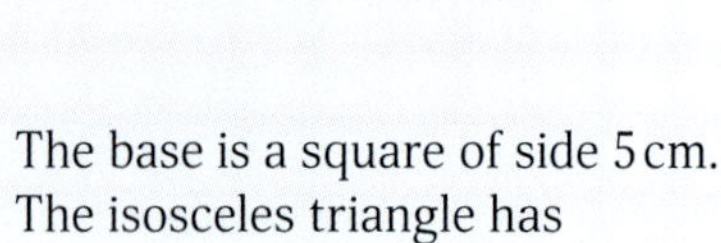

The base is a square of side 5 cm. The isosceles triangle has measurements as shown.

2 Tetrahedron

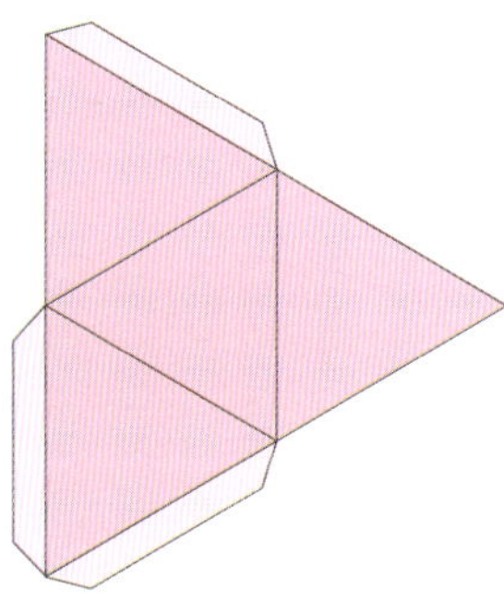

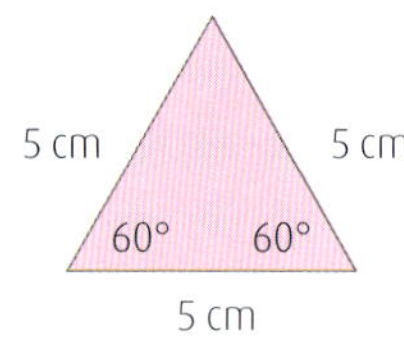

Each equilateral triangle has measurements as shown.

3 Triangular prism

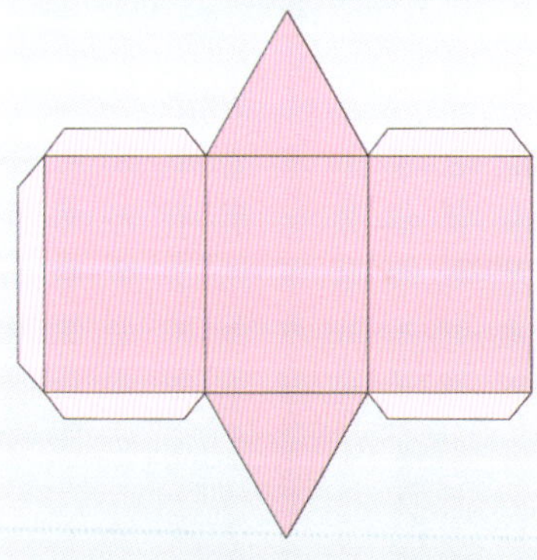

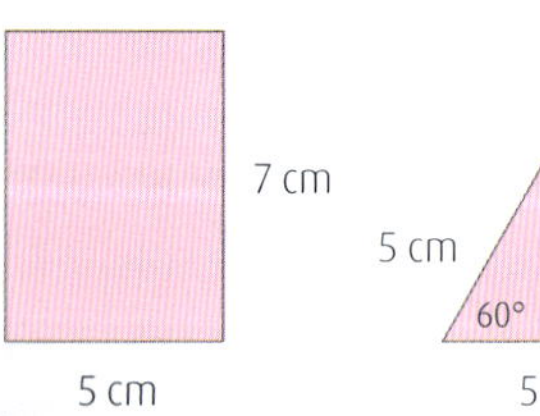

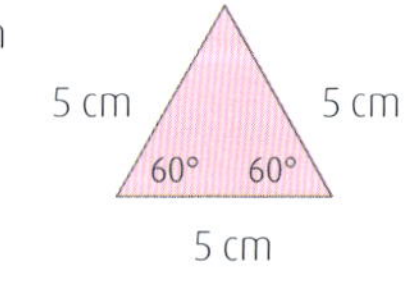

Each rectangle has measurements as shown. Each equilateral triangle has measurements as shown.

Find an extension activity to make a dodecagon (which can be used as a calendar!) in the Teaching Notes on the Bright Red website.

ANGLES, SYMMETRY AND TRANSFORMATIONS

ANGLES

I can name angles and find their sizes using my knowledge of the properties of a range of 2D shapes and the angle properties associated with intersecting and parallel lines.
MTH 3-17a

What's coming up?

This Outcome and Experience will give you the opportunity to:

- name angles using mathematical notation, for example, ∠ABC
- identify corresponding, alternate and vertically opposite angles and use this knowledge to calculate missing angles
- use the angle properties of triangles and quadrilaterals to find missing angles.

What you already know

You have already learned how to:

- ✔ use mathematical language including acute, obtuse, straight and reflex to describe and classify a range of angles identified within shapes in the environment
- ✔ measure and draw a range of angles to within ±2°
- ✔ know that complementary angles add up to 90° and supplementary angles add up to 180° and use this knowledge to calculate missing angles.

Naming angles

An angle is constructed using two 'arms' that meet at a **vertex**.

How does that work?

Here we have two lines, or arms, AB and AC, that meet at vertex A.
We name this angle using the three letters.
The vertex must be the middle letter.
So this could be written as angle CAB or angle BAC.
The sign ∠ is used to indicate an angle.

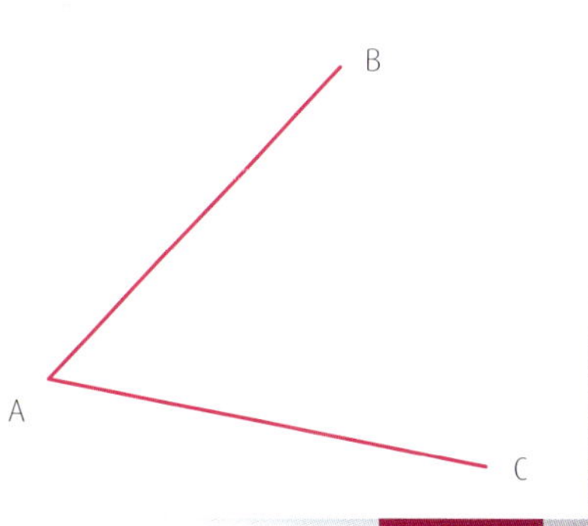

This angle would be written as ∠CAB or ∠BAC.
In some texts, only the letter at the vertex is used, for example here it could be written ∠A, but you should practise using three letters.

Classroom challenge

1 Use three letters to name each of these angles in two ways, as in the example above.

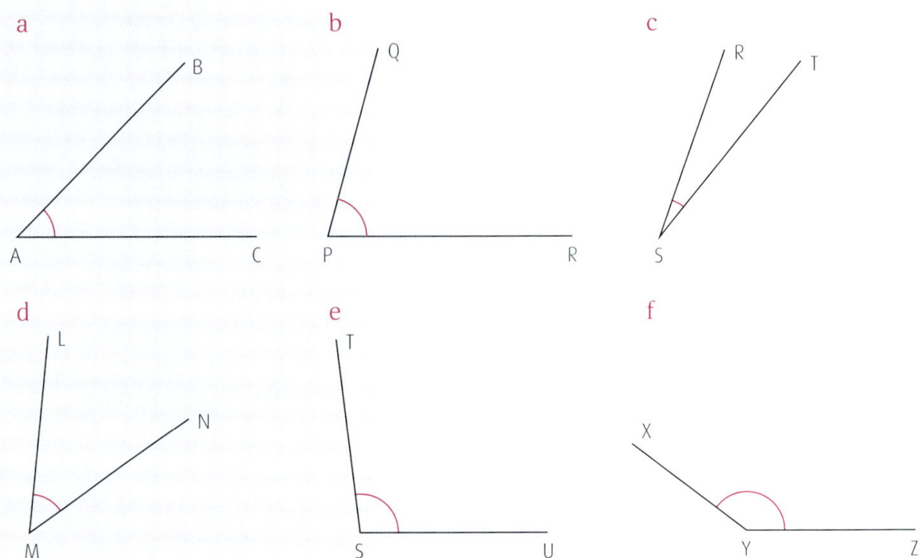

2 In each of these diagrams, name the two angles that share a vertex.
The first one is done as an example.

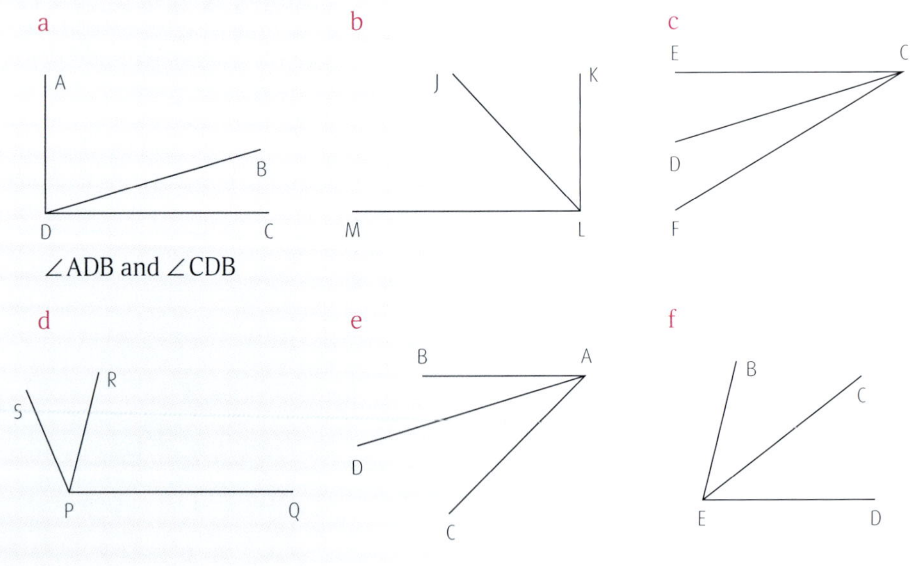

∠ADB and ∠CDB

3 Name all three angles in this diagram.

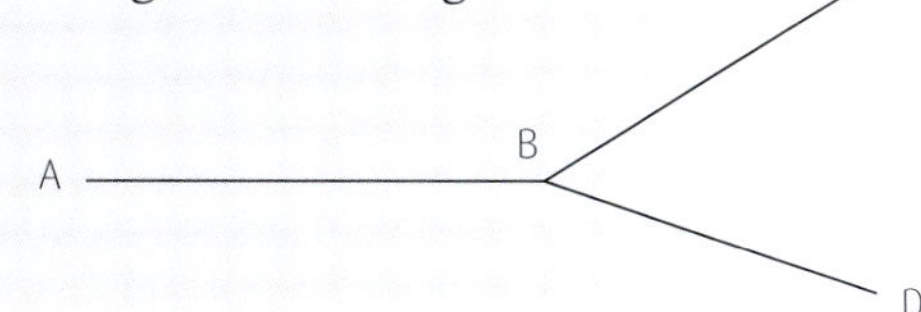

Parallel lines cut by a third line

When a line **intersects**, or crosses, a pair of parallel lines, a number of angles are formed.

In this section we will look at these angles and identify equal pairs of angles.

How does that work?

Look at this pair of parallel lines cut by a third line.
The arrows indicate that the lines are parallel.
To help you to identify pairs of equal angles you can look for 'shapes'.

In diagram 1, $a = b$, $c = d$, $e = f$ and $g = h$.
These angles form an 'X' shape.
X-angles are equal.
The proper name for X-angles is **vertically opposite** angles.

In diagram 2, $a = c$, $b = d$, $f = h$, and $e = g$.
These angles form an 'F' shape. The 'F' can be forwards, backwards or upside down!

For example:

2

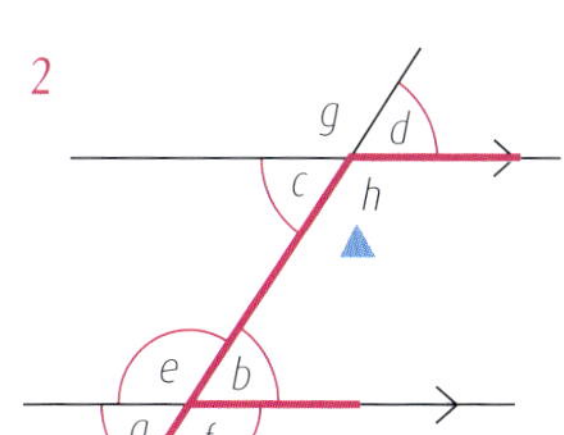

F-angles are equal.
The proper name for F-angles is **corresponding** angles.
They are in **corresponding** positions in the top and bottom set of angles.

In the diagram on the right, $c = b$ and $h = e$. The angles form a 'Z' shape. The 'Z' shape can be forwards or backwards.
Z-angles are equal.
The proper name for Z-angles is **alternate** angles.
They are on **alternate** sides of the line that cuts the pair of parallel lines.
We can use these equal pairs to calculate missing angles in diagrams.

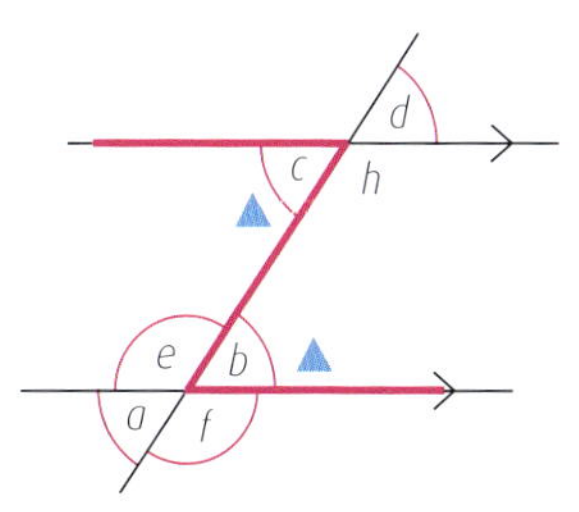

How does that work?

EXAMPLE

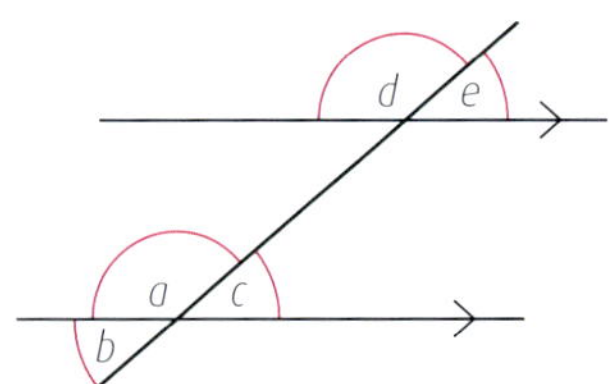

Use the angle facts on p 165 to calculate the sizes of the lettered angles if angle $a = 140°$.

SOLUTION

To find b:

$a + b = 180°$ They form a straight line so are **supplementary**, that is, they add up to 180°.

$140 + b = 180°$

$b = 40°$

To find c:

$c = b$ They form an 'X' shape so they are **vertically opposite** angles.

$c = 40°$

To find d:

$d = a$ They form an 'F' shape so they are **corresponding** angles.

$d = 140°$

To find e:

$e = c$ They form an 'F' shape so they are **corresponding** angles.

$e = 40°$

The other missing angles could be calculated in the same way.

DON'T FORGET

Reminder: complementary angles add up to 90°

Classroom challenge

1 Calculate the size of the angles marked with a letter.

a

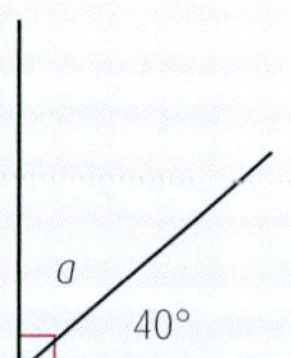

b

55°
b

c

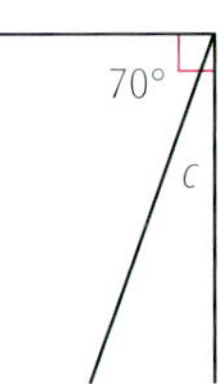

2 Calculate the size of the angles marked with a letter.

a

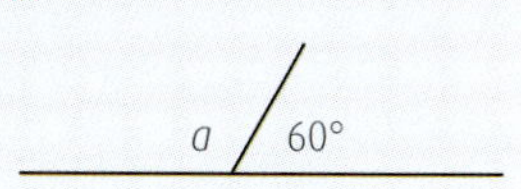

b

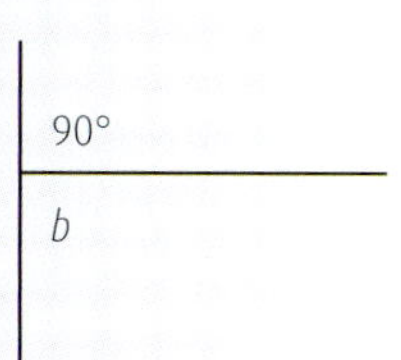

c

DON'T FORGET

Reminder: supplementary angles add up to 180°.

3 State the size of the angles marked with a letter.

a

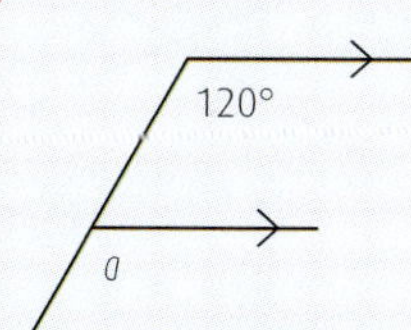

c

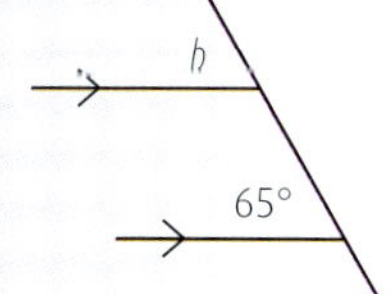

d

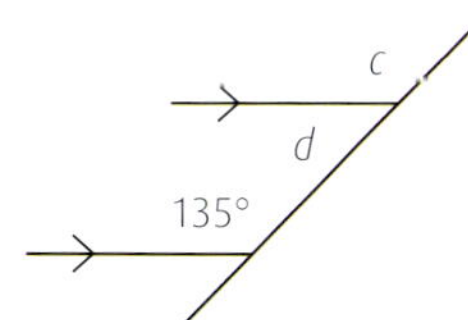

b What type of angles are $a°$ and 120°?
Copy and complete:
$a°$ and 120° are .... angles.

4 State the size of the angles marked with a letter.

a

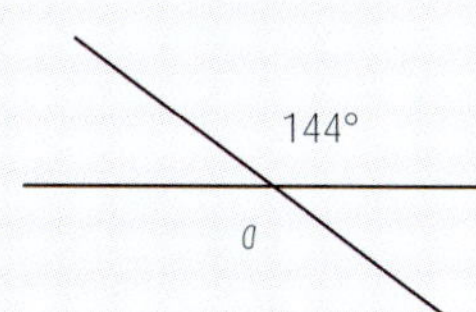

c

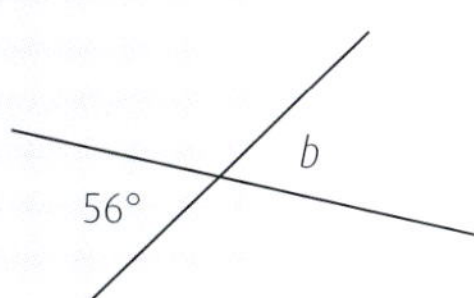

d

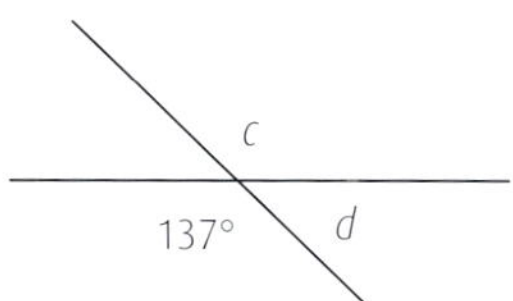

b What type of angles are $a°$ and 144°?
Copy and complete:
$a°$ and 144° are .... angles.

5 State the size of the angles marked with a letter.

a

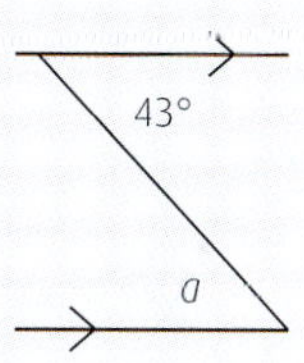

c

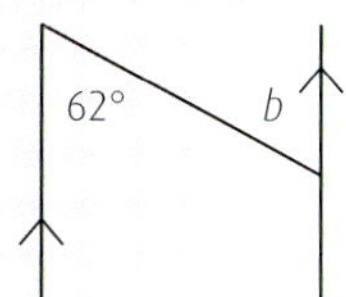

d

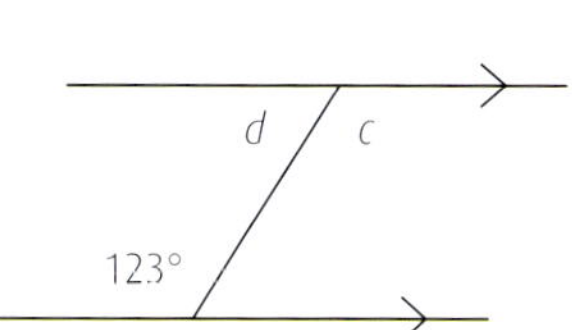

b What type of angles are $a°$ and 43°?
Copy and complete:
$a°$ and 43° are .... angles.

6 Using your knowledge of 'F', 'Z' and 'X' angles, calculate the size of the angles marked in these diagrams.
Try to write your answers like this:
Angle a is 44° as it is an **alternate** angle.
Angle b is 123° as it is a **vertically opposite** angle.
Note that you can often calculate an angle in different ways: you may say it is alternate, but your friend may have chosen to use vertically opposite angles.

a

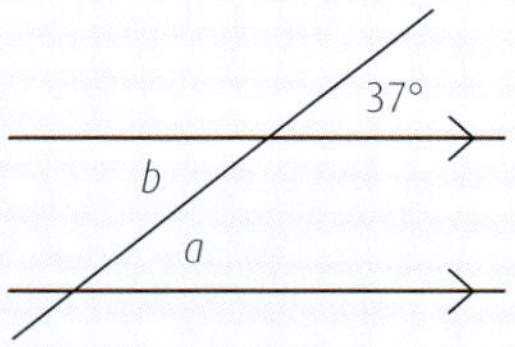

b

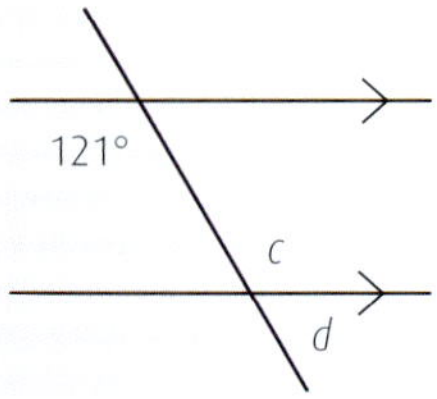

c

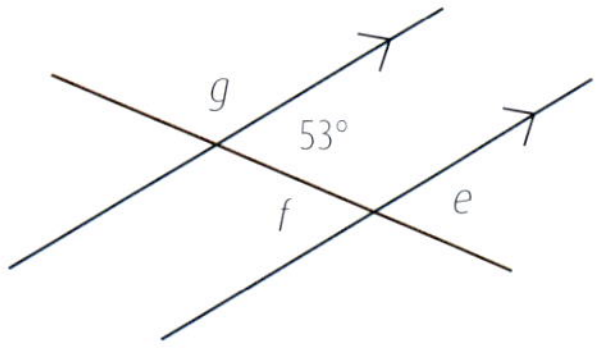

d

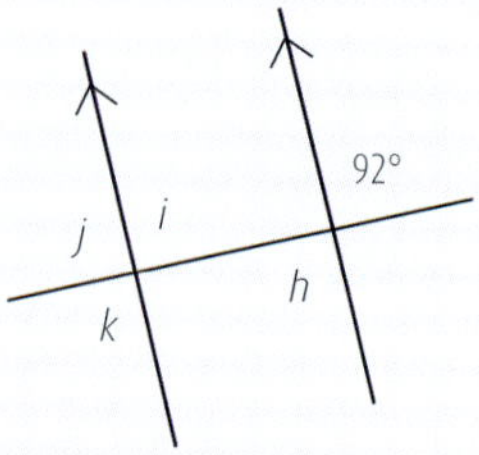

e

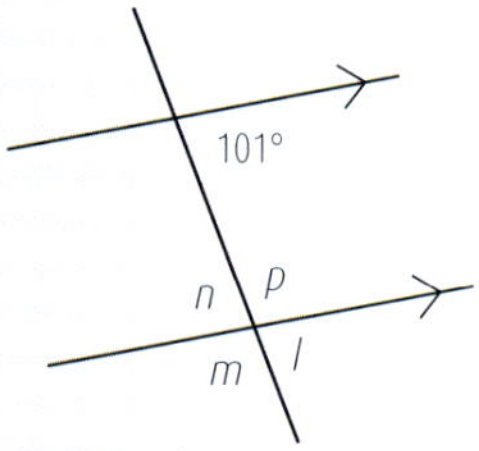

f

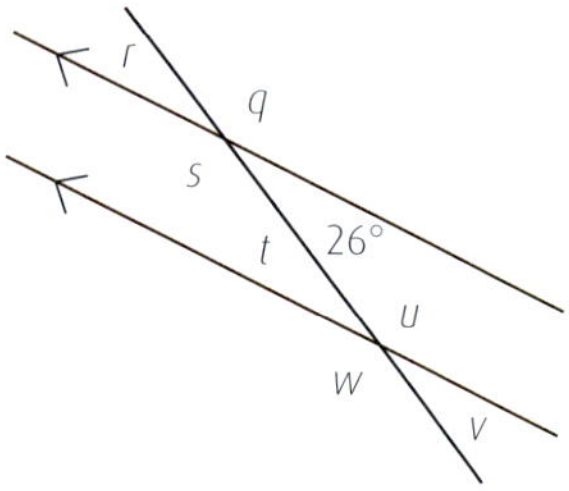

g

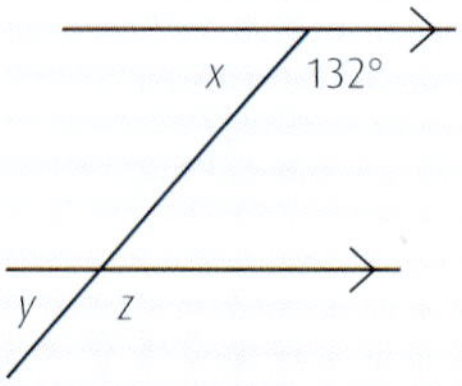

STRETCH YOURSELF

h

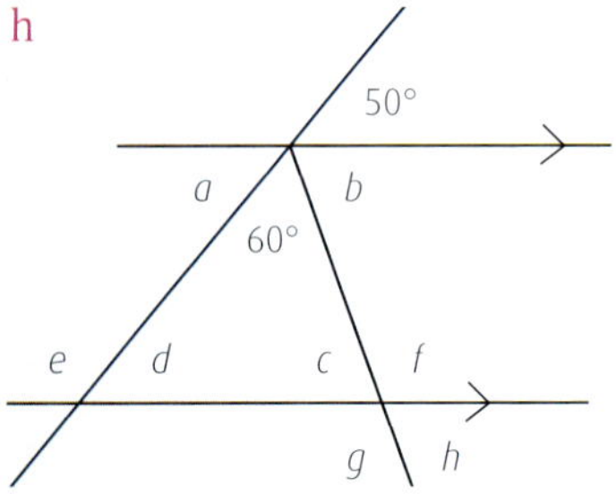

STRETCH YOURSELF

7 In the diagram, find the size of each angle marked with a letter.
Write down a reason for each answer.
For example, 'These two angles are corresponding.'
What is the name given to the type of triangle in the diagram?

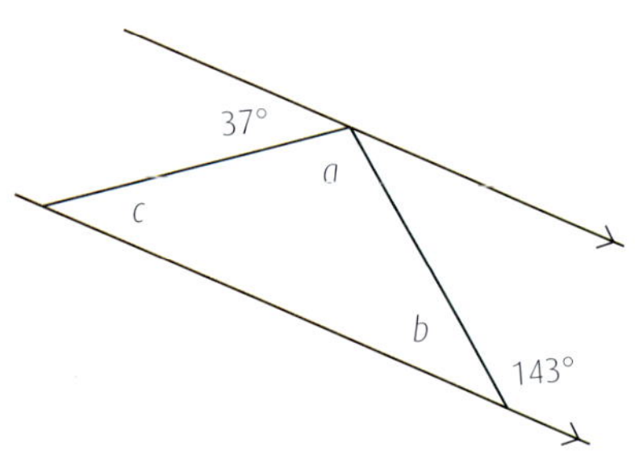

Angles in a triangle

The angles in a triangle add up to 180°.

How does that work?

Try this experiment.

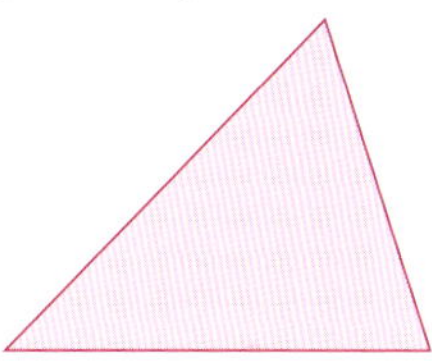

Draw a triangle.

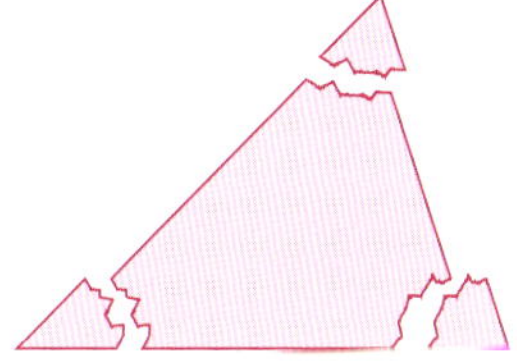

Tear off each of the corners.

Place them against a ruler.

The three angles will fit perfectly along the ruler's edge.
A straight angle is 180°.
Therefore, the three angles in a triangle must add up to 180°.
We can use this fact to calculate angles in a triangle.

EXAMPLE

Calculate the size of angle ABC in this triangle.

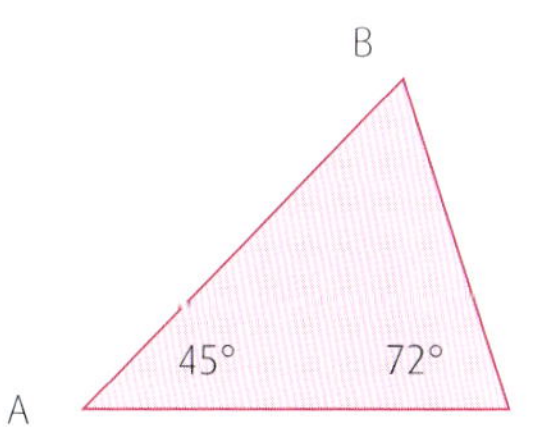

SOLUTION

Step 1 Add the known angles.

$$\begin{array}{r} 45 \\ +\ 72 \\ \hline =\ 117° \\ \hline \end{array}$$

Step 2 Take this total away from 180.

$$\begin{array}{r} 180 \\ -\ 117 \\ \hline =\ 63° \\ \hline \end{array}$$

∠ABC = 63°
Check by adding: 45 + 72 + 63 = 180°

Classroom challenge

1 In each of these triangles, calculate the size of the angle marked with a letter.

a

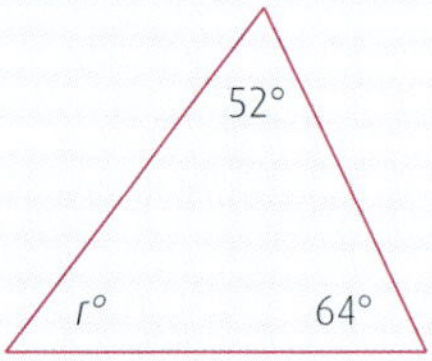

52°

r°

64°

b

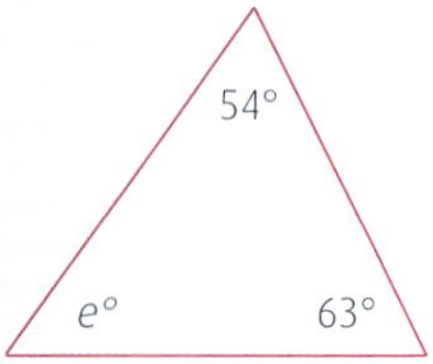

54°

e°

63°

c

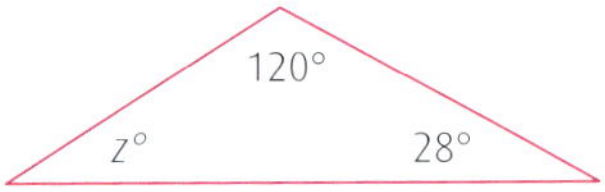

120°

z°

28°

d

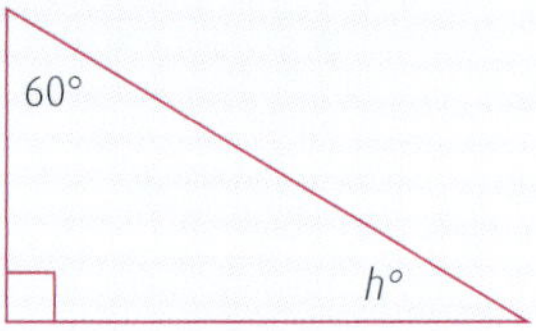

60°

h°

e

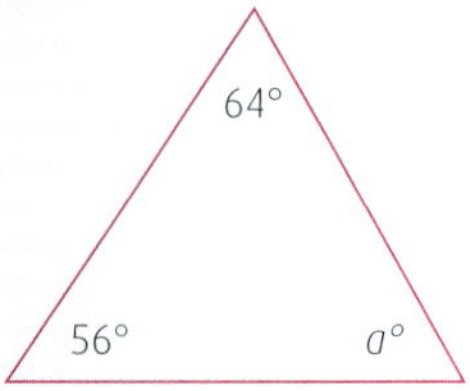

64°

56°

a°

f

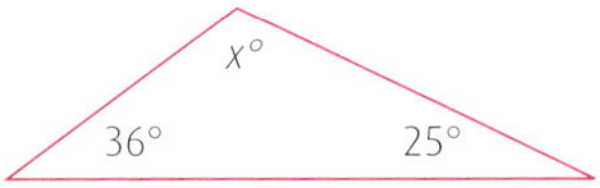

x°

36°

25°

g

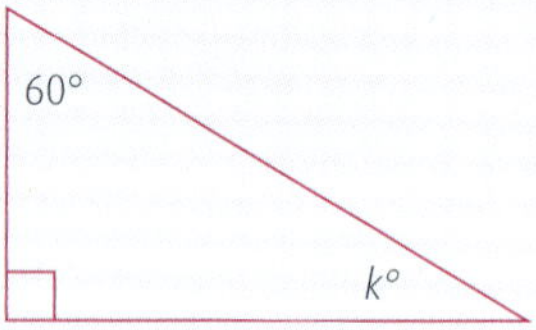

60°

k°

h

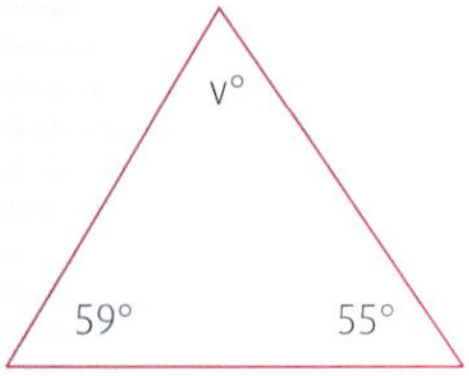

v°

59°

55°

i

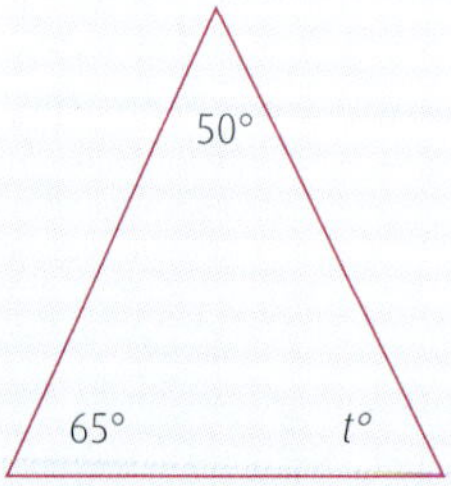

50°

65°

t°

What name is given to this type of triangle?

j

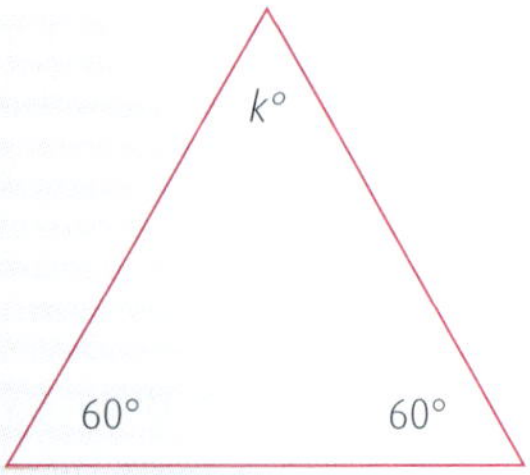

k°

60°

60°

What name is given to this type of triangle?

2 In these isosceles triangles, calculate the size of the missing angles.

a

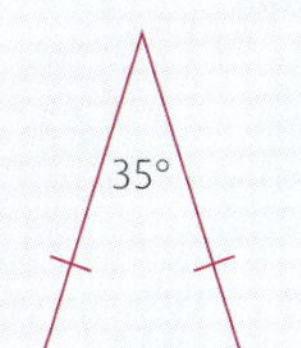

b

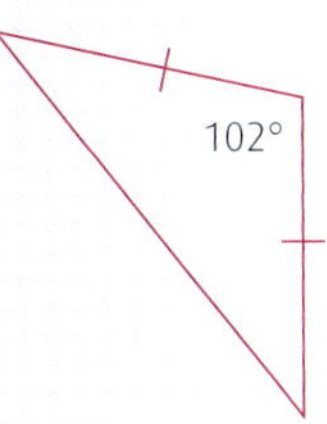

c

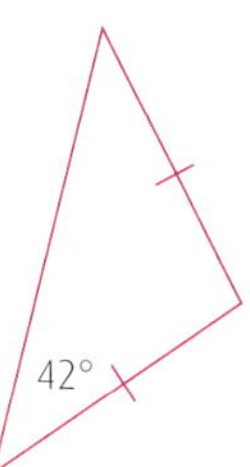

3 Calculate the size of each angle marked with a letter.

a

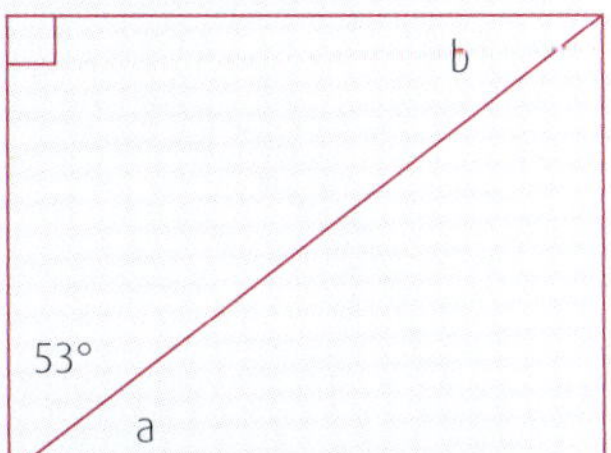

b

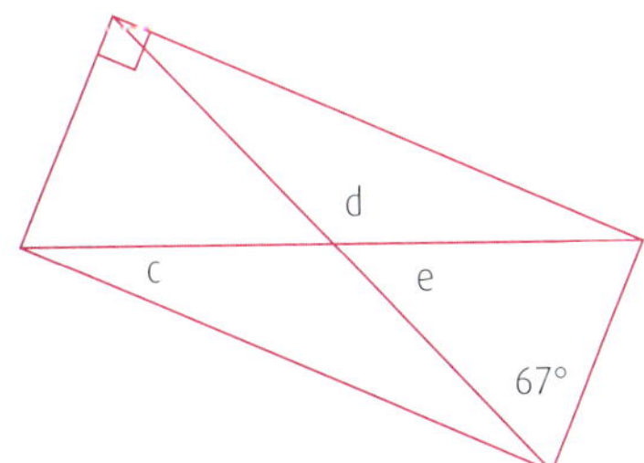

Angles in a quadrilateral

Here is a quadrilateral.

What do you think the four angles will add up to?

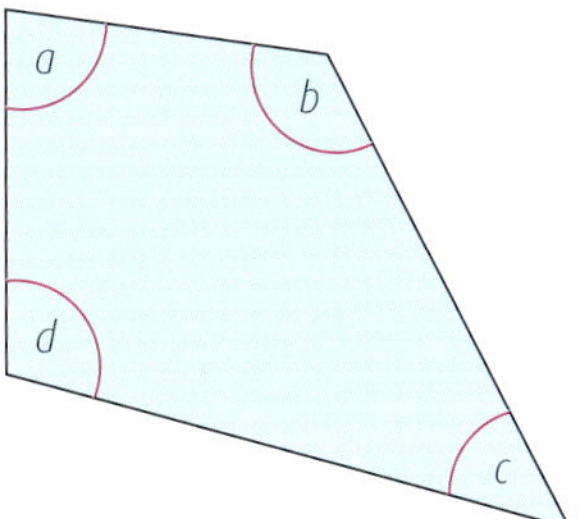

How does that work?

Try an experiment like you did for the triangle.
Tear off the four corners.
Try to fit them together.

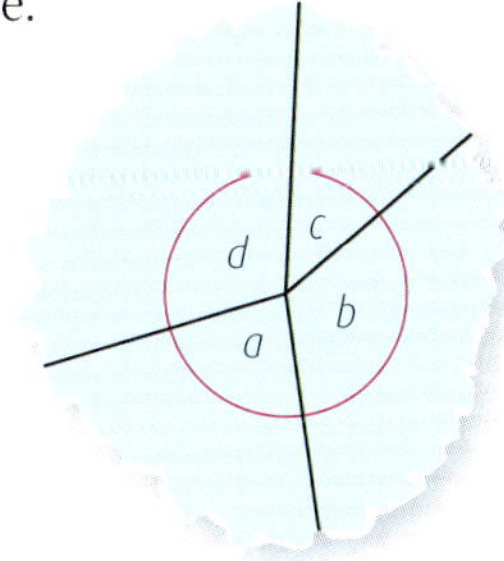

What happens?
Did you see that the four angles form a complete circle?
That means that the four angles in a quadrilateral add up to **360°**.

Classroom challenge

1 Use this fact to calculate the size of the angle marked x in each of these quadrilaterals.

BEARINGS AND SCALE

Bearings

Having investigated navigation in the world, I can apply my understanding of bearings and scale to interpret maps and plans and create accurate plans, and scale drawings of routes and journeys.
MTH 3-17b

What's coming up?

This Outcome and Experience will give you the opportunity to:

- use bearings in a navigational context, including creating scale drawings.

What you already know

You have already learned how to:

✔ use your knowledge of the link between the eight compass points and angles to describe, follow and record directions.

DON'T FORGET

A bearing is a measure of direction, taken from a starting point of facing north.

How does that work?

Bearings are measured clockwise from north.
3 figures are used for bearings.
North has a bearing of 000°.
The diagram shows an 8-point compass, with bearings marked.

In this diagram, if you walk from O, in the direction of the arrow, you are walking on a bearing of110°.

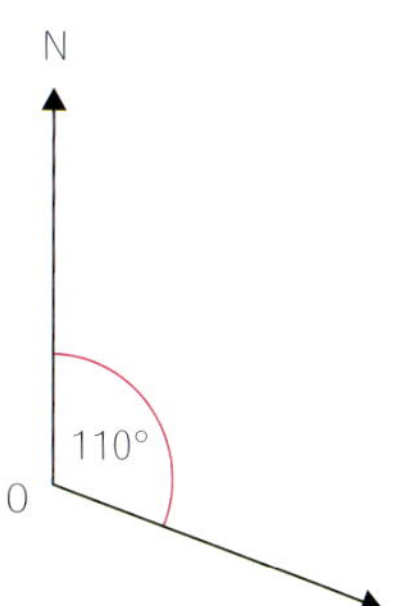

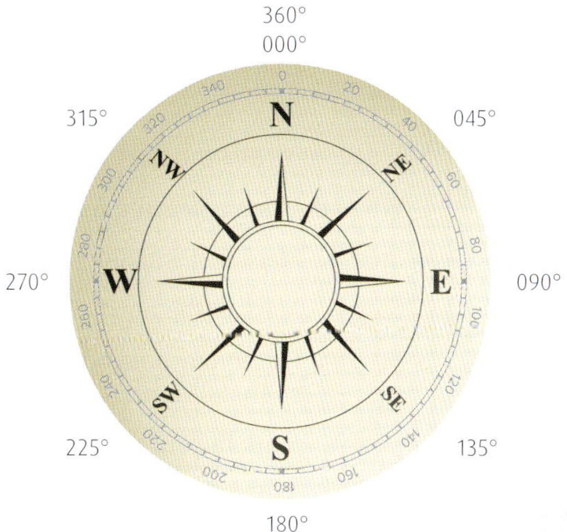

Here are some more examples of bearings.

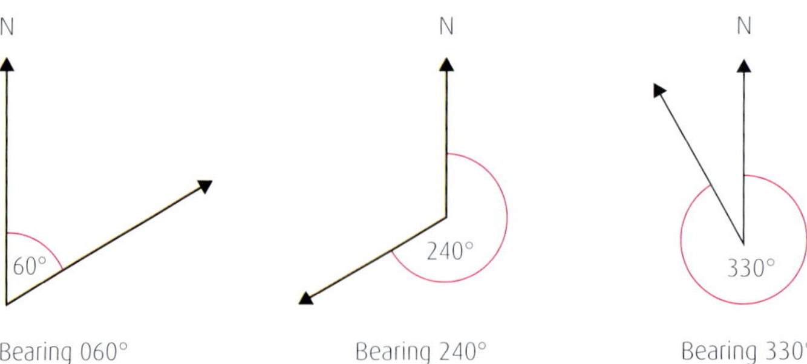

Bearing 060°   Bearing 240°   Bearing 330°

EXAMPLE

A ship is sailing from A to B on a bearing of 060°, as shown in the diagram.

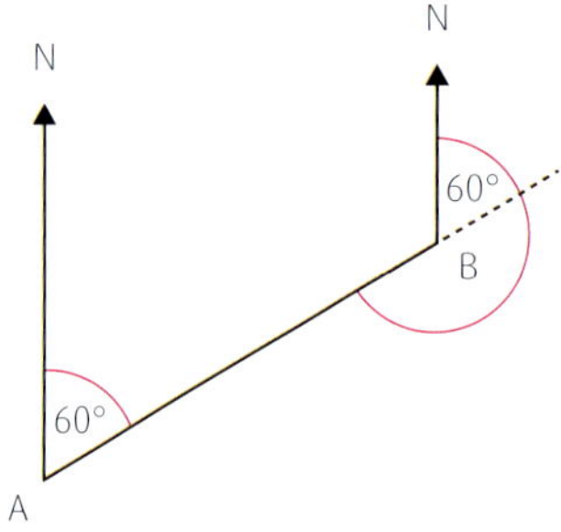

On what bearing must it sail to return from B to A?

SOLUTION

Step 1 Extend the 'bearing' line to create parallel lines cut by a third line.

Step 2 Mark in the corresponding angles ('F' shape).

Step 3 Add 180° (straight line).
60 + 180

Step 4 Write this bearing down.
The bearing of A from B is 240°.

How to draw a bearing

When drawing a bearing follow these steps.

Step 1 Mark in a starting point.

Step 2 Mark in the north arrow.

Step 3 Place your protractor on starting point so that 0° is on the north arrow.

Step 4 Mark in your required angle.

Step 5 Join the starting point to this mark.

EXAMPLE 1

From point P, draw a bearing of 165°.

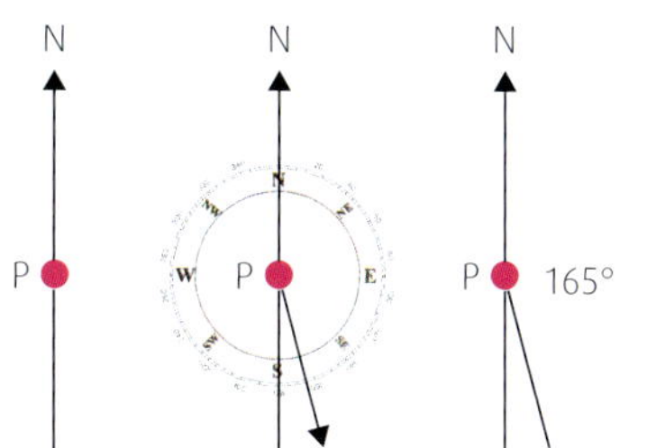

SOLUTION 1

Step 1 Draw a north arrow through point P.

Step 2 Place your protractor on point P with 0 lying along north line.

Step 3 Make a mark at 165°.

Step 4 Join point P to this mark (and extend if required).

EXAMPLE 2

You use the same process to measure the bearing of one point from another.

What is the bearing of Q from P?

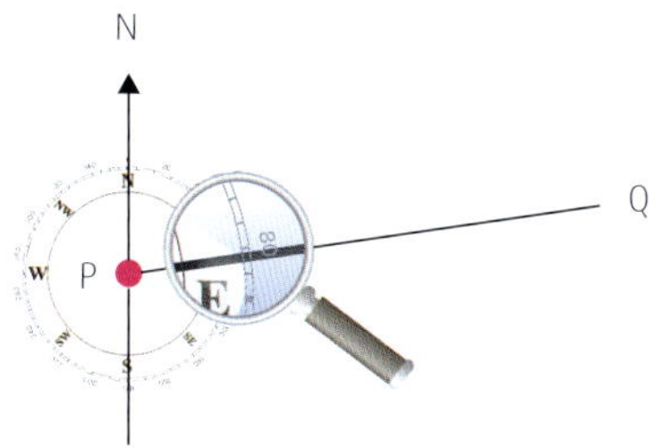

SOLUTION 2

Step 1 Draw a north line through point P.

Step 2 Place your protractor on point P with 0 lying along north line.

Step 3 Read the angle from your protractor.

Step 4 write this as a 3-figure bearing of 082°.

A 360-degree protractor is excellent for working with bearings.

However, a 180-degree protractor can be used just as easily.

To draw a bearing of, say, 195°, use the following steps.

Step 1 Mark in your starting point and the north line.

Step 2 Lightly extend the north line through your starting point.

Step 3 North–south has used up 180°.

Step 4 Set your protractor as shown and measure round 195 – 180 = 15 degrees.

Step 5 Mark in the point and join with a line.

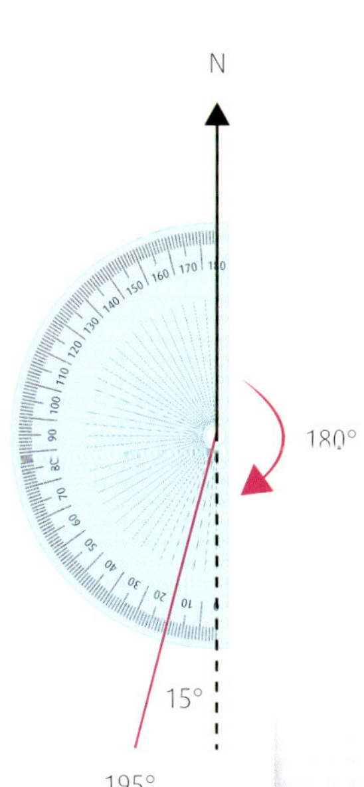

This is now showing a bearing of 195°.

The same technique would be used for measuring such a bearing.

You know that north–south is 180°.

Simply measure the 'extra' angle and add on 180° to get the bearing.

1 The map of an island is shown below.

From the Waterfall, write down the bearing of:

a the harbour
b the mine
c the church
d the landing strip
e the lighthouse
f the tower.

2 The diagram shows the position of two ships at sea.

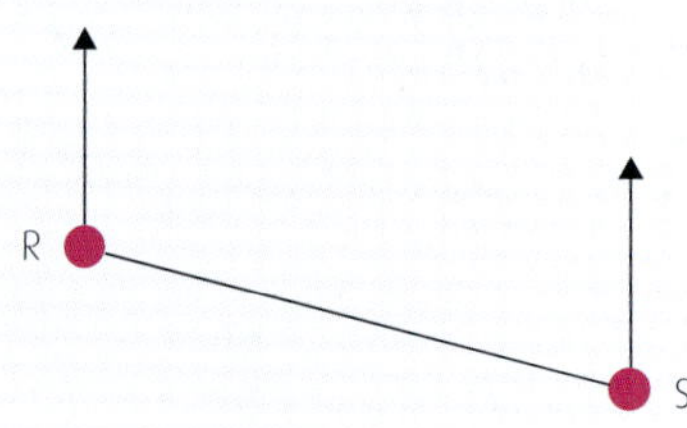

a What is the bearing of ship S from ship R?
b What is the bearing of ship R from ship S?

3 The diagram shows the position of three towns, Aytown, Beetown and Ceetown.

a What is the bearing of B from A?
b What is the bearing of C from B?
c What is the bearing of C from A?
d What is the bearing of B from C?
e What is the bearing of A from B?

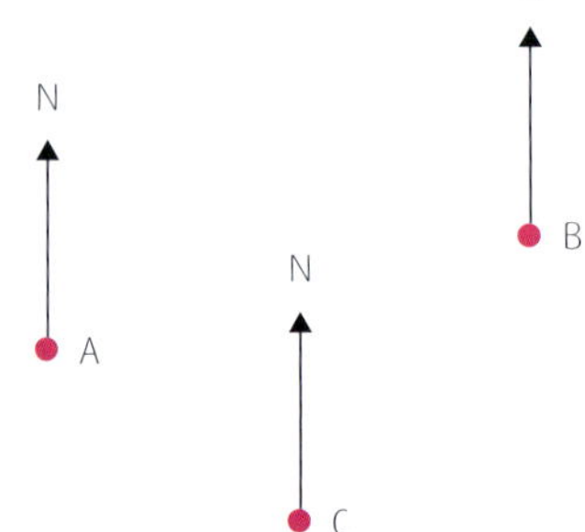

4 Four 'EasyAir' planes leave the airport at Exeter and fly to four different destinations. The flights are shown in the table below.

Flight	Destination	Bearing
EA001	London	078°
EA002	Glasgow	355°
EA003	Leeds	037°
EA004	Guernsey	162°

Copy and complete this diagram to show the direction in which each plane flew to its destination.

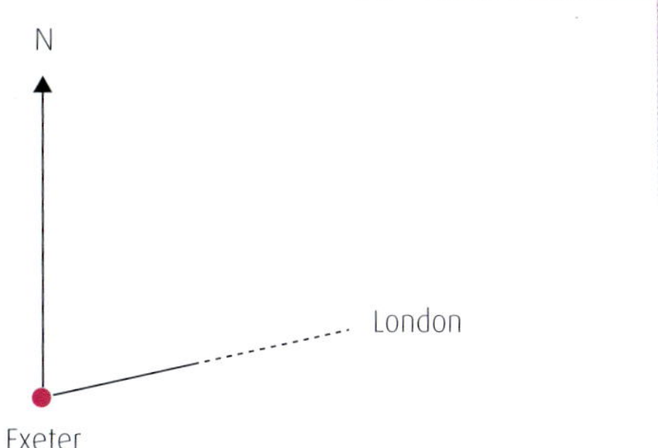

5 Here is a map of part of Scotland.

Use the map to find the bearing of:

a Aberdeen from Fort William
b Ullapool from Braemar
c Wick from Oban
d Perth from Jedburgh
e Ayr from Dundee
f Inverness from Ayr
g Edinburgh from Glasgow
h Braemar from Aviemore
i Dumfries from Inverness
j Stirling from Oban
k Dundee from Uig
l Stranraer from Jedburgh
m Jedburgh from Wick
n Wick from Glasgow.

Using scale and bearings

We can use scale drawings, along with bearings, to solve navigation problems, or to plan a route.

How does that work?

EXAMPLE

A ship sails from a port P for 35 kilometres on a bearing of 060°.
It then changes course and heads due south for 25 kilometres.
It then stops.

a How far is the ship from its starting point?

b On what bearing must it sail to return to the port?

SOLUTION

Use a scale of 1 cm represents 5 km.
The path of the ship can now be accurately drawn.

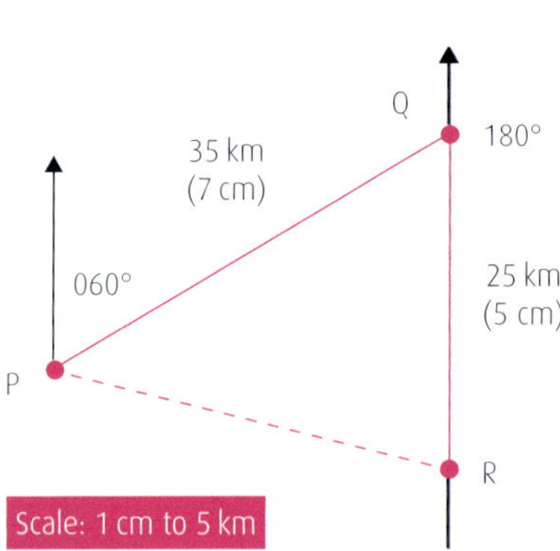

35 ÷ 5 = 7 → 7 cm 1st leg

25 ÷ 5 = 5 → 5 cm 2nd leg

a The distance RP measured in centimetres = 6·2 cm
For each 1 cm, the 'real' distance is 5 km. So the real distance is 6·2 × 5 = 31 km.

b The bearing of P from R is measured as 285°.

EXAMPLE

Claire walks for 750 metres on a bearing of 030°.
She then walks on a bearing of 315° until she is due north of her starting point.
Claire then stops for a break.

a How far does Claire walk on the bearing of 315°?

b How far is Claire from her starting point when she stops for her break?

SOLUTION

There is enough information to construct a scale drawing.
A suitable scale to use would be 1 cm to 100 m.

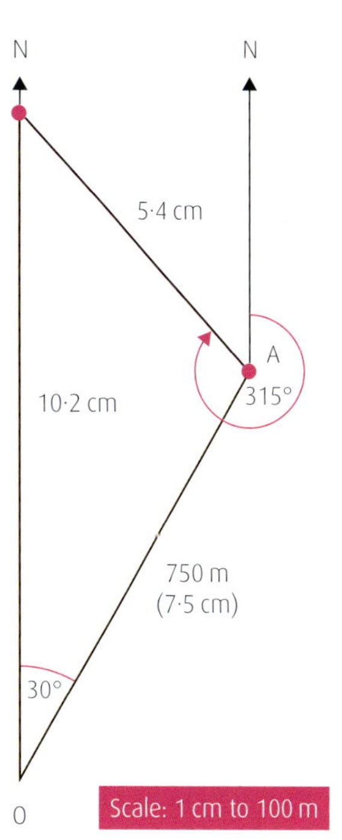

DON'T FORGET

Choose a scale which will give you a 'decent' size diagram.

750 ÷ 100 = 7·5 so first line will be drawn 7·5 cm long.

a Second line is measured as 5·4 cm.
5·4 × 100 = 540 So second leg of journey is 540 metres long.
Claire walked 540 m on the bearing of 315°.

b Distance from start is measured as 10·2 cm.
10·2 × 100 = 1020
Distance from start is 1020 metres.

Classroom challenge

1 Sonia walks 120 metres due north, then 180 metres due east.
  a Use a scale of 1 cm to 20 m to construct a scale drawing of Sonia's journey.
  b How far will Sonia have to walk to get back to the start?
  c On what bearing should Sonia walk to get back to the start?

2 James walks 600 metres north-east and then walks 800 metres south and stops for a rest.
  a Use a scale of 1 cm to 200 m to construct a scale drawing of James's journey.
  b If James had gone directly from the start to his resting place, on what bearing should he have gone?
  c How far would he have walked if he had taken this direct route?

3 An aeroplane flies for 300 km on a bearing of 065°.
It then flies on a bearing of 290° until it is due north of its starting point.
  a Use a suitable scale (e.g. 1 cm to represent 50 km) to construct a scale drawing of the plane's journey.
  b How far is the plane from its starting position?

4 Arthur intended to walk 200 m on a bearing of 195°.
However, his compass was not working properly, and he actually walked 200 m on a bearing of 215°.
What distance, and on what bearing, should Arthur walk to get to his intended finishing point?

5 The diagram shows a map of an island.
P and Q are two houses on the island.

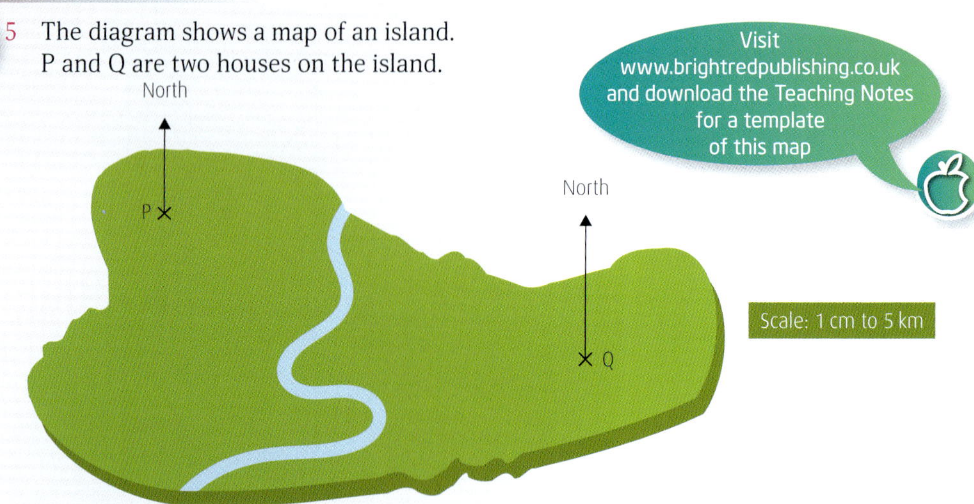

a What is the bearing of P from Q?

b How far apart, in kilometres, are the two houses?

c Another house, R, is 20 km from P on a bearing of 130°.
Mark this house on the map of the island.

d A fourth house is on a bearing of 270° from Q and on a bearing of 175° from P.
Mark this house on the map with an S.

6 The diagram shows an island with north lines drawn at two points, A and B.

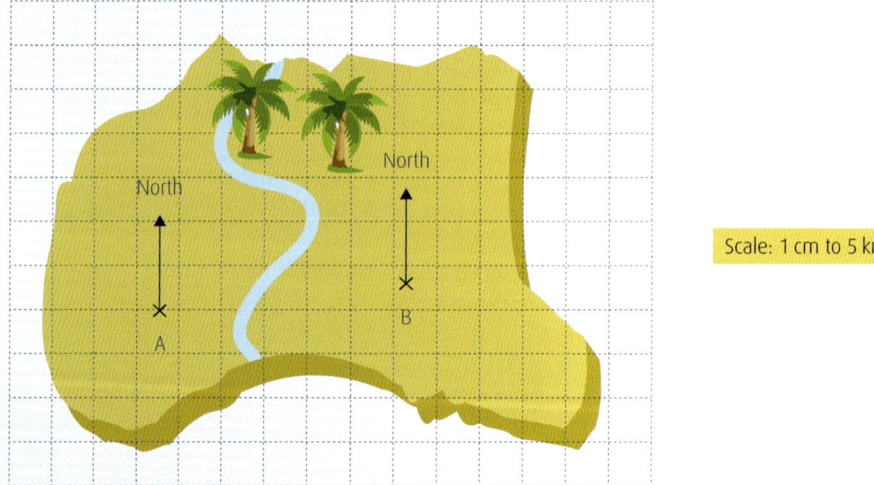

a Calculate the real distance between points A and B.

b A treasure is buried on the island!
It is buried on a bearing of 038° from point A and on a bearing of 285° from B.
Mark the position of the treasure with an 'X'.

Visit
www.brightredpublishing.co.uk
and download the Teaching Notes
for a template
of this map

7 The diagram shows part of a coastline.
Port B is 52 km due north of port A.
Craig wants to sail his yacht from port A to port B.
He cannot go directly.
Craig sets out from port A on a bearing of 075° and sails for 28 km.
He then changes course and sails for another 24 km on a bearing of 335°.
He then drops anchor to check his position.

a Using a suitable scale (e.g. 1 cm to represent 8 km), construct a scale drawing of Craig's route.

b How far is Craig from port B?

c On what bearing should Craig now sail to arrive at port B?

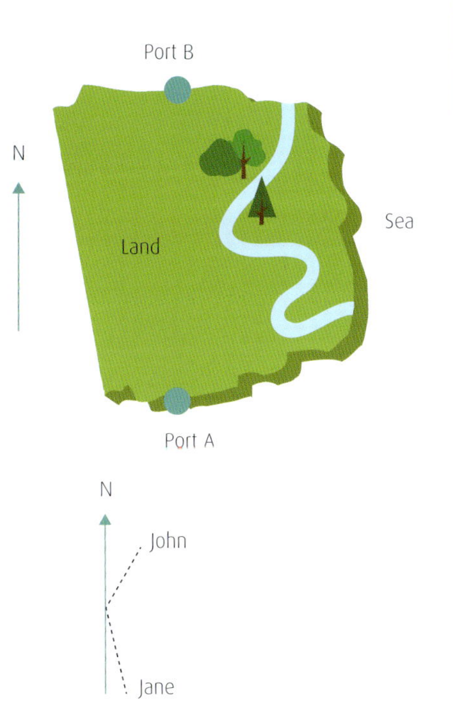

8 John and Jane set off walking from the same point.
John walks 300 m on a bearing of 034°.
Jane walks 310 m on a bearing of 165°.
How far apart are they now?

9 An aeroplane flies for 150 km on a bearing of 330°.
It then changes course and flies for 100 km on a bearing of 175°.
It then flies for 350 km on a bearing of 260° after which it lands.

a Construct a scale drawing of the plane's journey.

b How far is the plane from the starting point of the journey?

c On what bearing could the plane have flown to go directly from start to finish?

How does that work?

EXAMPLE

The table shows the distances and directions of four airports from Exeter Airport.

Destination	Distance (km)	Bearing
London	250	078°
Glasgow	580	355°
Leeds	390	037°
Guernsey	150	162°

Using a suitable scale (e.g. 1 cm to represent 50 km), construct a map showing the position of all five airports.

Sometimes, the scale is given as a **ratio** or a **representative fraction**.

A map has a scale of 1 : 50 000.

This means for each unit on the map, there are 50 000 units 'on land'.

That is:

1 cm on map → 50 000 cm actual size
= 500 m

The distance between two houses on the map is 4·5 cm.

How far apart are the houses?

1 cm → 50 000 cm
4·5 cm → 50 000 × 4·5
= 225 000 cm
= 2250 m
= 2·25 km

Classroom challenge

1 A map has a scale of 1 : 50 000.
On the map, the distance between two churches is 3 cm.
How far apart, in kilometres, are the churches?

2 A map is drawn to a scale of 1:50 000.
On the map, the distance between the train station and the shopping centre is 3·7 cm.
How far away, in kilometres, is the shopping centre from the train station?

3 A map is drawn to a scale of 1 : 100 000.
On the map, two towns are 8 cm apart.
How far apart, in kilometres, are the two towns?

4 A map is drawn to a scale of 1 : 150 000.
On the map, the airport is 7·2 cm from the town centre.
How far away, in kilometres, is the airport from the town centre?

5 Here is a map of Midlothian. It is drawn to a scale of 1 : 200 000.

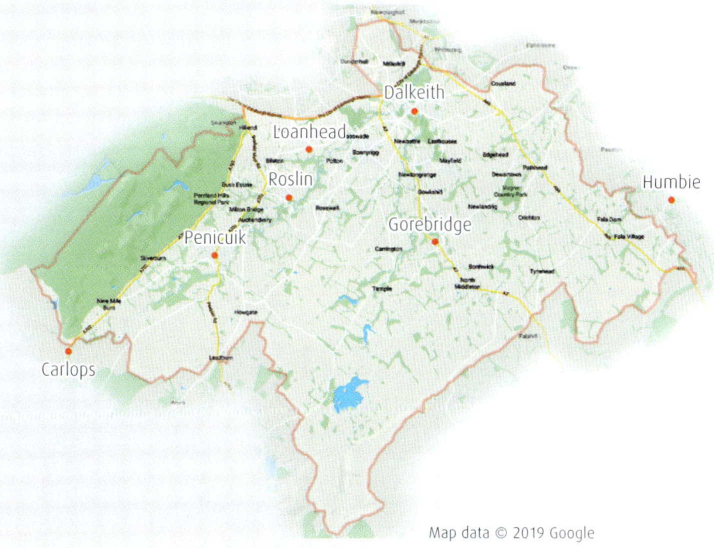

Map data © 2019 Google

a What is the bearing and distance of Gorebridge from Dalkeith?
b What is the bearing and distance of Penicuik from Humbie?
c What is the bearing and distance of Loanhead from Gorebridge?
d What is the bearing and direction of Roslin from Carlops?
e What is the bearing and distance of Gorebridge from Penicuik?

6 A map is drawn to a scale of 1 : 50 000.
Two houses are 3 kilometres apart.
How far apart, in centimetres, would they be on the map?

DON'T FORGET

Copy and complete

3 km = 3000 m
= 300 000 cm
(convert to unit to be used on map)
Scale: 1 cm → 50 000 cm
300 000 ÷ 50 000 = ? cm
The towns would be .... cm apart on the map.

7 A map is drawn to a scale of 1 : 100 000.
Two towns are 15 kilometres apart.
How far apart, in centimetres, would the towns be on the map?

8 A map is drawn to a scale of 1 : 150 000.
Two hilltops are 6·5 kilometres apart.
How far apart, in centimetres, would the hilltops be on the map?

9 A map is drawn to a scale of 1 : 2000.
A garden plot measures 70 metres in length by 30 metres broad.
a What would the length of the plot be, in centimetres, on the map?
b What would the width of the plot be, in centimetres, on the map?

10 A 'stellar map' is drawn to a scale of 1 : 2 000 000 000 000.
The approximate distance from the Earth to the Sun is 150 000 000 kilometres.

a How far apart, in centimetres, would the Earth and Sun be on the map?

b On the map, two asteroids are 2·5 centimetres apart.
How far apart, in kilometres, are the two asteroids?

STRETCH YOURSELF

11 A 'stellar map' is drawn to a scale of 1 : 10 000 000 000 000.
The table shows the distances, in kilometres, of each of the planets (or dwarf planets) in our Solar System from the Sun.

a Copy the table and complete the column to show the distance on the map, in centimetres, of each planet from the Sun.

Planet (or Dwarf planet)	Distance from Sun (km)	Distance on map (cm)
Mercury	58 000 000	0·58 cm
Venus	108 000 000	
Earth	150 000 000	
Mars	228 000 000	
Jupiter	780 000 000	
Saturn	1 430 000 000	
Uranus	2 870 000 000	
Neptune	4 500 000 000	
Pluto (Dwarf Planet)	6 000 000 000	

b Draw a line 60 centimetres long. At one end of the line mark a dot and name it 'The Sun'.
Now mark each planet with a dot (with the name underneath) to show the relative distance of each planet from the Sun.

12 Here is a map of the State of Victoria, in Australia.

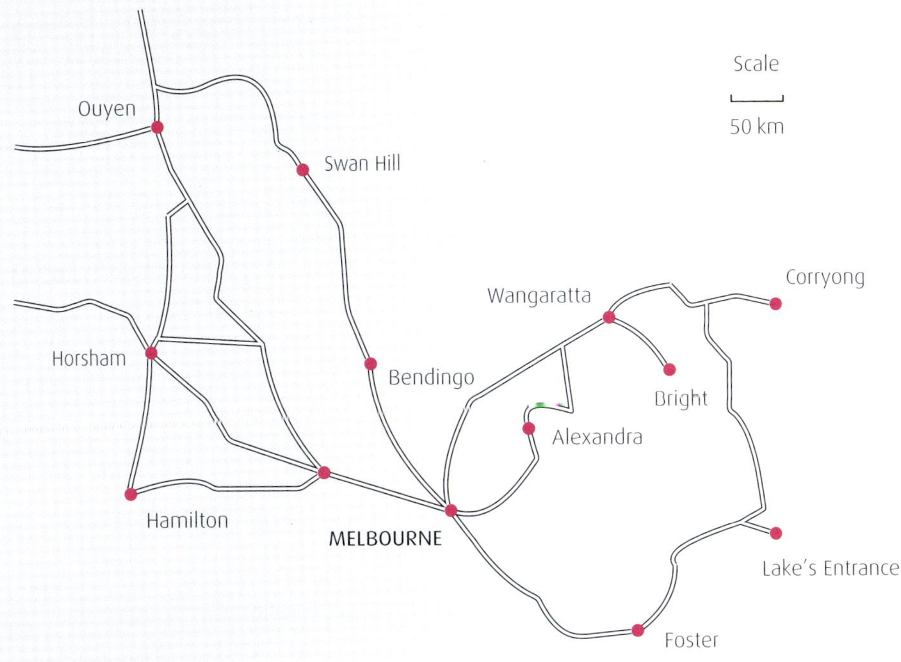

Using the scale given, 1 cm to 50 km, or 1:5 000 000, measure and then calculate the real, direct, distance between:

a Bendingo and Melbourne
b Ouyen and Bright
c Hamilton and Wangaratta
d Corryong and Lake's Entrance
e Swan Hill and Foster
f Horsham and Alexandra.

By using thin string, and laying it along the roads, can you calculate more accurately the distances between the towns above?

Scale drawings and plans

I can apply my understanding of scale when enlarging or reducing pictures and shapes, using different methods, including technology. MTH 3-17c

What's coming up?

This Outcome and Experience will give you the opportunity to:

- apply knowledge and understanding of scale to enlarge and reduce objects in size showing understanding of linear scale factor.

What you already know

You have already learned how to:

- ✔ interpret maps, models or plans with simple scales, for example, 1 cm : 2 km.

Can you think where you may have used scale drawings or plans?

Have you ever helped to put together flat-pack furniture (for example, from IKEA)?

Have you ever made a model aeroplane from a kit (for example, an 'Airfix' model)?

Have you ever looked at a picture, taken through a microscope, of a microbe?

Have you tried to redesign your bedroom, by seeing what furniture may fit in?

Have you looked at a photograph of a monument (for example, the Eiffel Tower)?

Have you looked at a diagram of a circuit board for a computer?

All of these involve plans or images of reductions or enlargements of an item.

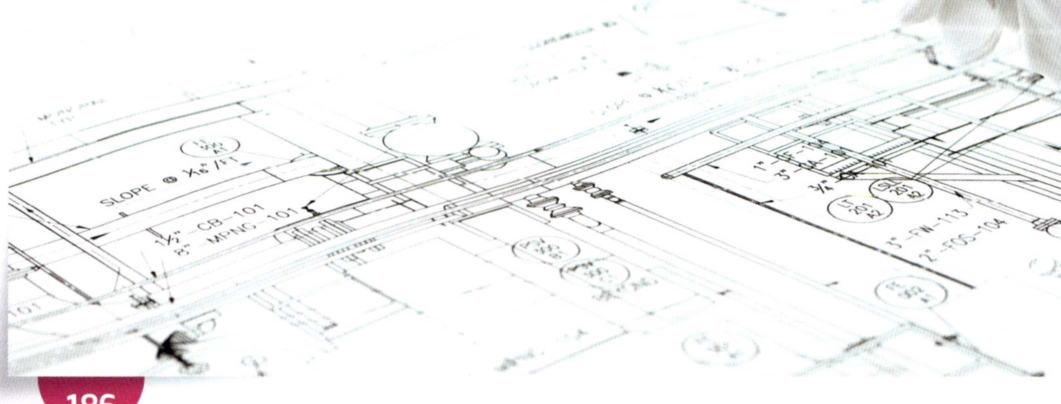

Enlarge or reduce shapes

Shapes can be **enlarged** or **reduced** by a scale factor.

How does that work?

For example, if you are asked to draw an **enlargement** of a shape using a scale factor 3 simply **multiply** all the lengths by 3.

This rectangle would become

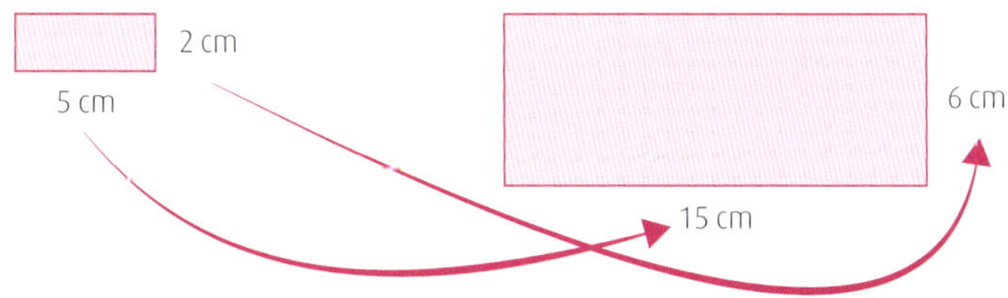

In the same way, for a **reduction** of the rectangle using a scale factor of $\frac{1}{2}$, simply **multiply all the lengths by $\frac{1}{2}$.**

This rectangle

would become

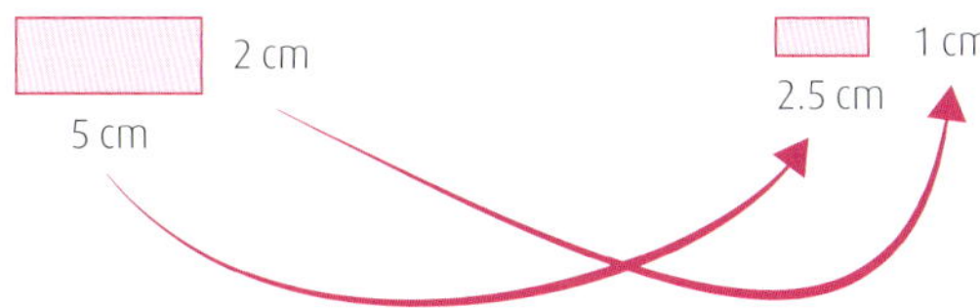

Classroom challenge

1 Draw **enlargements** of these shapes using a **scale factor of 2**.
Each square on the grid is 1 cm by 1 cm.

a

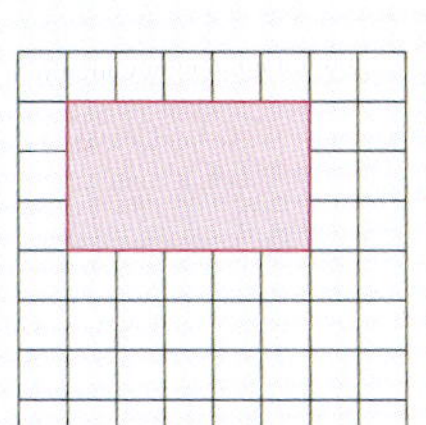

b

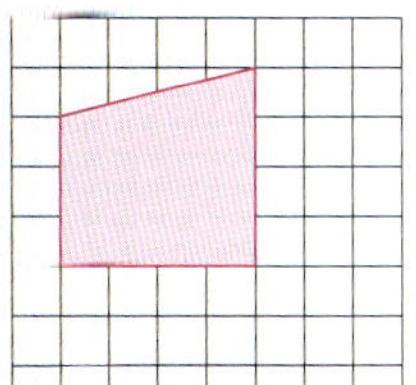

c

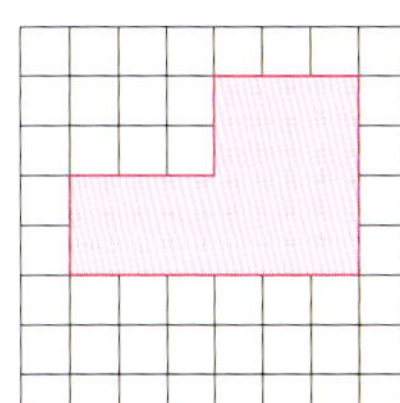

d

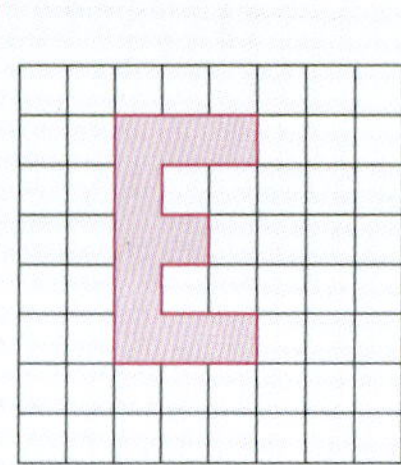

e

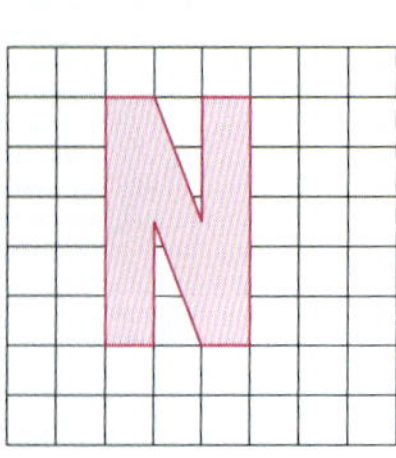

f

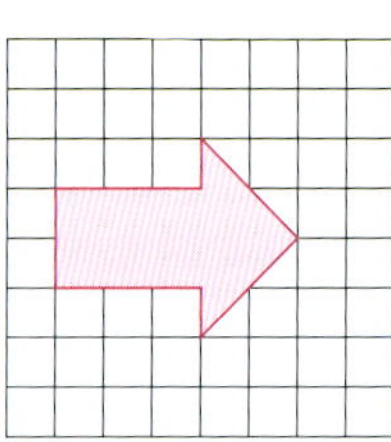

2 Draw **reductions** of these shapes using a **scale factor of** $\frac{1}{2}$. Each box is 1 cm by 1 cm.

a

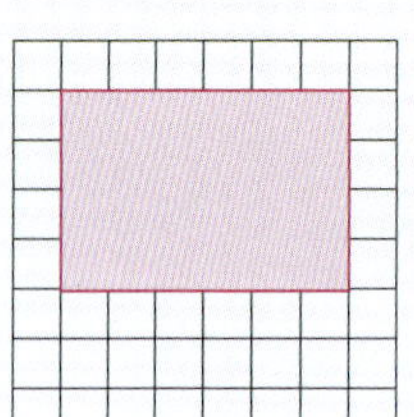

b

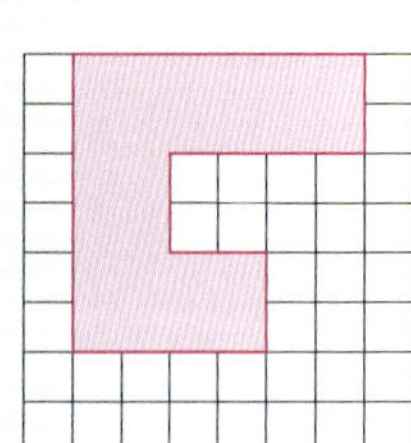

c

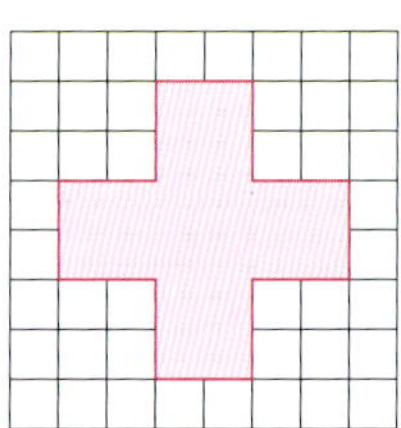

3 Draw a **reduction** of this flag of Russia. Use a **scale factor of** $\frac{1}{3}$. Each box is 1 cm by 1 cm.

4 Draw an **enlargement** of this triangle. Use a **scale factor of 3**.

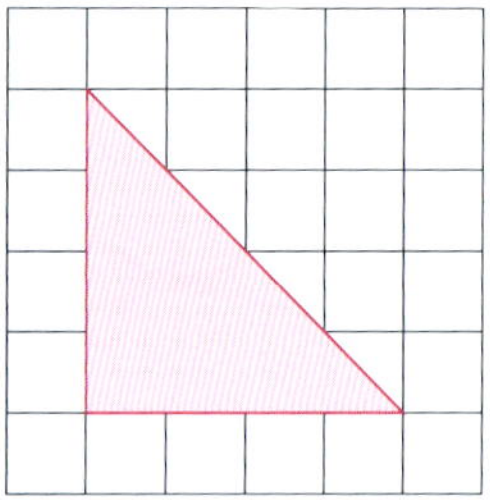

5 Draw **reductions** of these shapes. Use a **scale factor of** $\frac{1}{2}$.

a

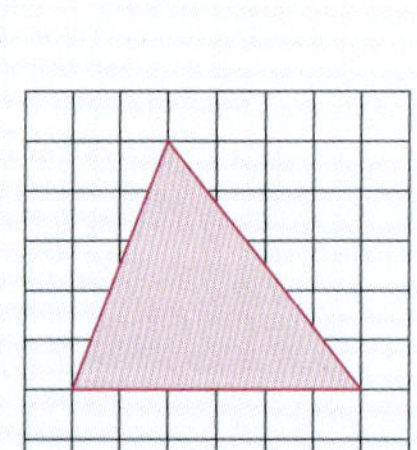

b

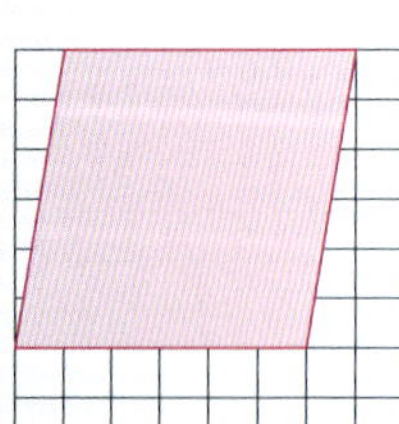

c

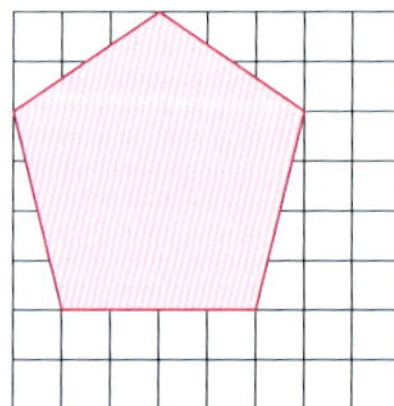

6 Draw a **reduction** of this house. Use a **scale factor of** $\frac{1}{4}$.

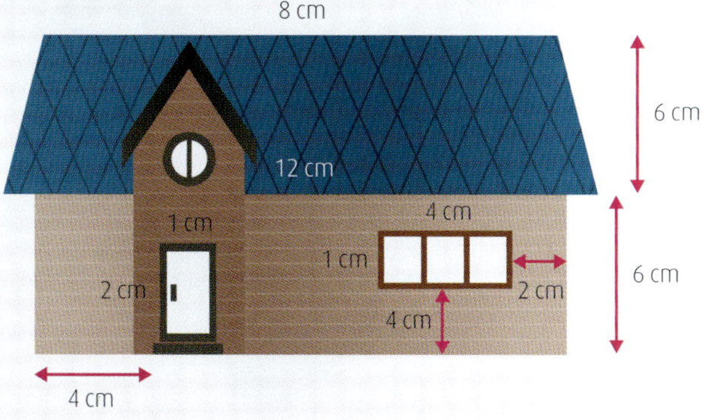

How does that work?

We can use scale factor to calculate real sizes from scale drawings, or scale sizes from real measurements.

EXAMPLE

The picture shows a model kit for the Spitfire plane. The kit is made to a scale of 1 : 24.

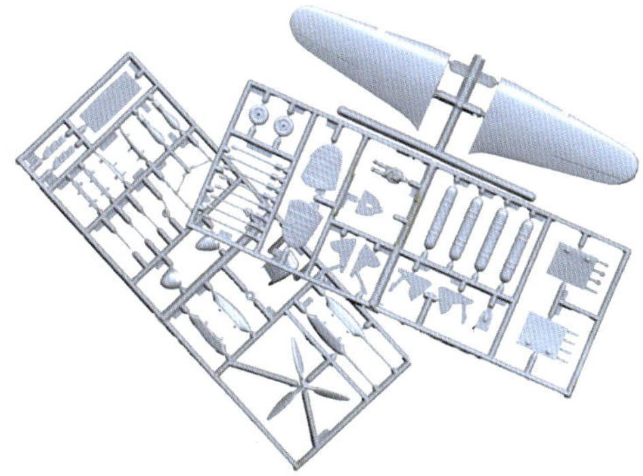

The wingspan of the model measures 46 cm. What is the wingspan of the real Spitfire?

SOLUTION

Scale		Real
1	→	24
46	→	**24** × 46
		= 1104 cm
		= 11·04 metres

The real wingspan is 11·04 metres.

EXAMPLE

The length of the model measures 38 cm. What is the length of the real Spitfire?

SOLUTION

Scale		Real
1	→	24
38	→	**24** × 38
		= 912 cm
		= 9·12 metres

The real length of the Spitfire is 9·12 metres.

EXAMPLE

Another model, of the same Spitfire is to be made to a scale of 1 : 72.
What will be the wingspan and length of this model?

SOLUTION

Real wingspan 11·04 metres = 1104 cm
Scale 1 : 72

Real	Scale
72	1
1	1 ÷ 72
1104	1 ÷ 72 × 1104
	= 15·3 cm

Model wingspan will be 15·3 centimetres.

Real length 9·12 metres = 912 cm
Scale 1 : 72

Real	Scale
72	1
1	1 ÷ 72
912	1 ÷ 72 × 912
	= 12·7 cm

Length of model will be 12·7 centimetres.

How does that work?

EXAMPLE

The diagram, in an instruction booklet, of a bookcase is drawn to a scale of 1 : 30.
The diagram shows the bookcase as:

7·5 cm high
5·4 cm wide
1 cm deep

What are the actual dimensions of the bookcase?

Height 7·5 × **30** = 225 cm or 2·25 m
Width 5·4 × **30** = 162 cm or 1·62 m
Depth 1 × **30** = 30 cm

A diagonal support strap on the bookcase is 2·80 metres long.
What will be the length of the strap when shown in the instruction manual?

Length 2·80 m = 280 cm 280 ÷ **30** = 9·3 cm

To calculate the scale factor, look at the corresponding sides in each figure.

How does that work?

EXAMPLE

The height of a man is 1·8 metres. His height in a photograph is 6 centimetres. What is the scale factor from man to photograph?

SOLUTION

Scale factor = Real : Scale
= 6 : 180
= 1 : 30

DON'T FORGET

Make sure you use the **same units**.

EXAMPLE

In the same photograph, a building is 13 cm high.
What is the real height of the building?

SOLUTION

Scale factor = 1 : 30

Set out like this: Scale : Real

	Scale	:	Real	
× 13	1	:	30	× 13
	13		30 × 13 = 390	

Real height of building = 390 cm = 3·9 metres

Classroom challenge

1 A real beluga whale measures 4·2 metres in length.
A sketch of the whale is drawn in a book with length as shown.

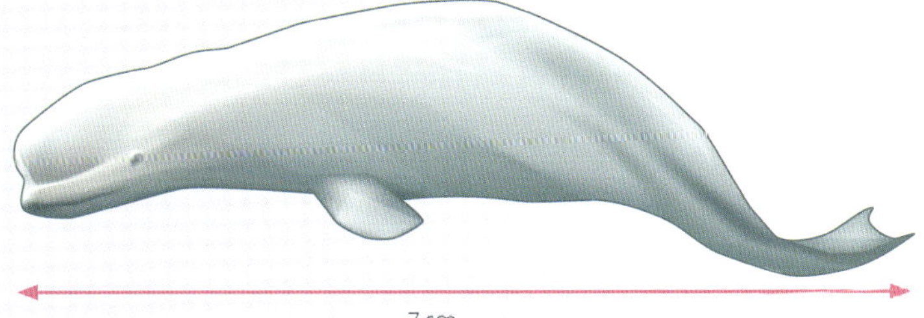

a What is the scale factor for reducing from real to picture?
Copy and complete. Scale factor = scale : real
= 7 : … remember use same units
= 1 : …

b A female great white shark measures 5·4 metres in length. Using the same scale, what will be the length of the sketch of the great white shark?

c In the same book, a sketch of a dolphin is 2·5 centimetres long. What is the real length of the dolphin?

2 This lorry has been drawn using a scale of 1 cm to 1·5 m.

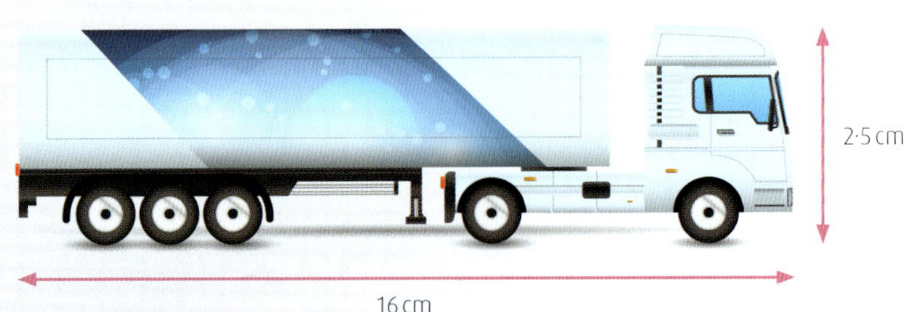

a Calculate the real length of the lorry.
b Calculate the real height of the lorry.

3 Jeremy has made a sketch of his garden. The real width of the garden is 15 metres.

a Calculate the scale factor that Jeremy used.
b Calculate the real length of the garden.
c Calculate the real area of the garden, in square metres.

4 Jamie's garage is 8 metres long and 3 metres wide.
What would be the length and width of Jamie's garage on drawings made to a scale of:

a 1:100 b 1:50 c 1:200 d 1:10?

5 On a scale drawing, a sideboard has dimensions as follows:
length 3·6 centimetres, width 1·2 centimetres and height 1·8 centimetres.
The real height of the sideboard is 90 cm.

a What was the scale factor used to draw the plan?
b What is the real length of the sideboard?
c What is the real width of the sideboard?

6 On a plan, drawn to a scale of 1:120, a rectangular room has a length of 4 cm and a width of 3 cm.

a What are the real length and width of the room?
b Jane thinks the diagonal distance from corner to corner is 6 metres.
Is Jane correct?

7 Here is a sketch of a kitchen.
Make an accurate scale drawing using a scale of 1 cm to 1 m.

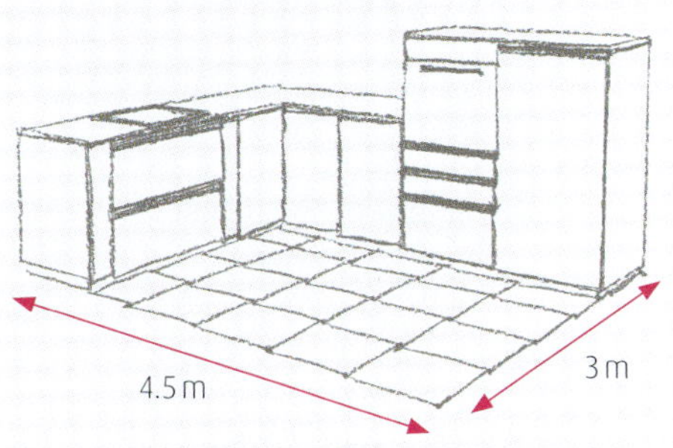

8 Here is a sketch of the end of a garden shed.
Make an accurate scale drawing using a scale of 1 : 25.

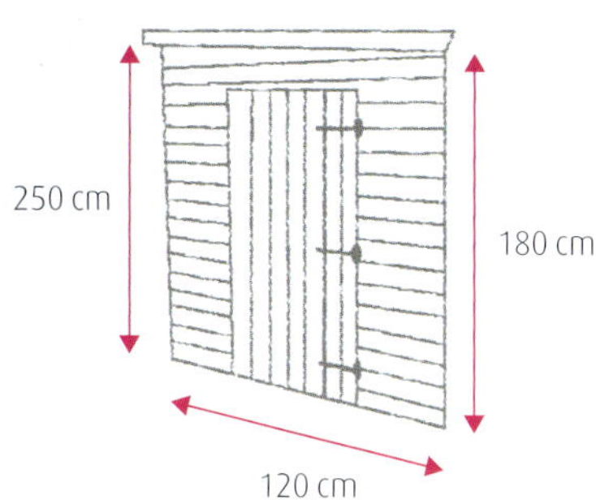

Using a protractor and ruler to make a scale drawing

Scale drawings are useful ways of finding the real heights of, for example, buildings.

How does that work?

EXAMPLE

Callum is trying to find the height of the Eiffel Tower.
He walks a distance of 50 metres from the base of the tower.
He measures the angle to the top of the tower to be 80°.
Callum makes a scale drawing to calculate the height of the Eiffel Tower.

Step 1 Choose a suitable scale, for example, 1 cm to 25 m.

Step 2 Draw the 'base line'. 50 ÷ 25 = 2 so draw a line 2 cm long.

Step 3 At one end, mark in an angle of 80°.

Step 4 Lightly draw a line until it is vertically above the other end of the base line.

Step 5 Join the end of base line to the angled line and measure.
If you are doing this example, you should get a measurement of 13 cm.

Step 6 Multiply by scale factor.
Real height = 13 × **25** = 325
The Eiffel Tower is 325 metres high.

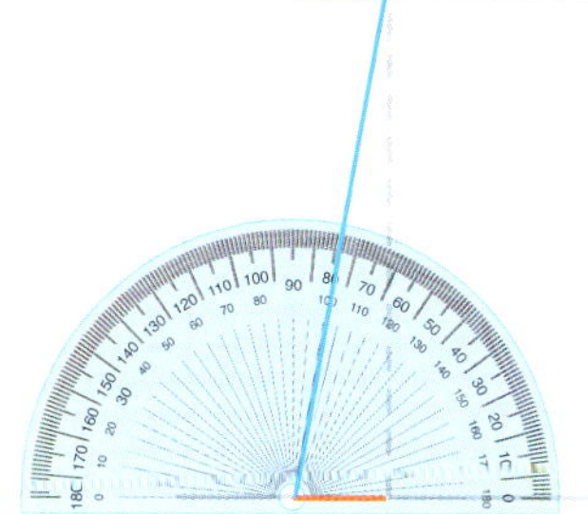

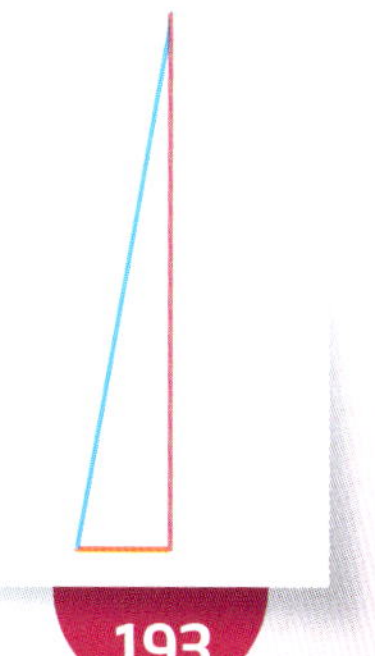

Classroom challenge

For each of the objects below:

a construct a scale drawing, using the scale given

b calculate the height of the object.

1 Use a scale of 1 cm to 2 m.
What is the height of the tree?

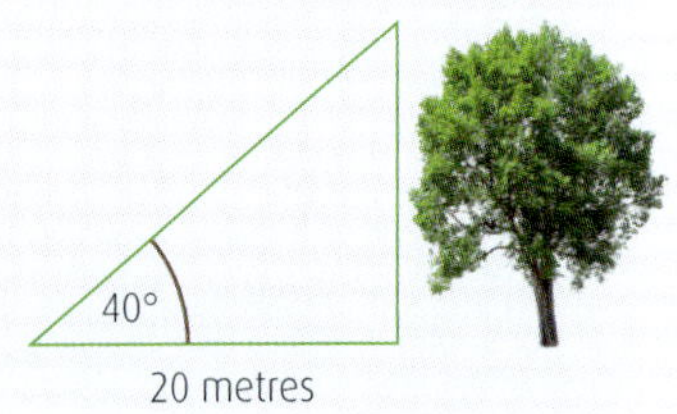

2 Use a scale of 1 cm to 3 m.

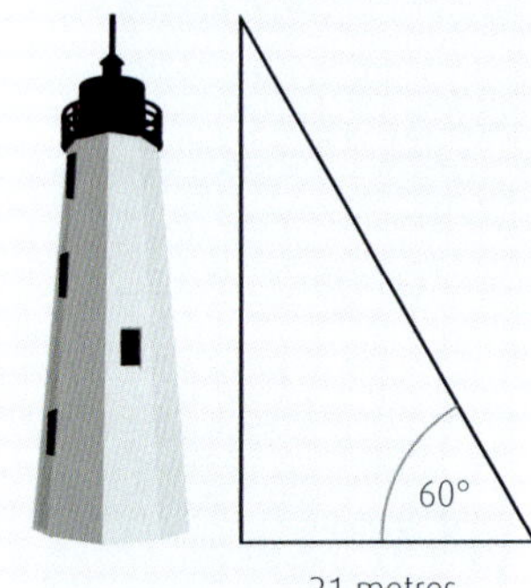

What is the height of the tower?

3 Use a scale of 1 : 500.
What is the height of the giraffe?

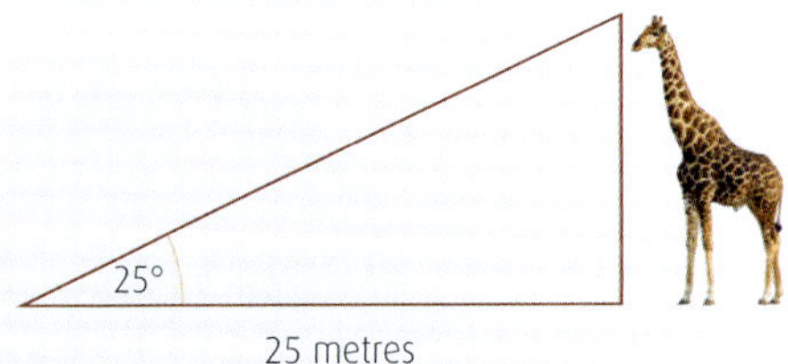

4 Use a scale of 1 : 5000.
What is the height of this rocket?

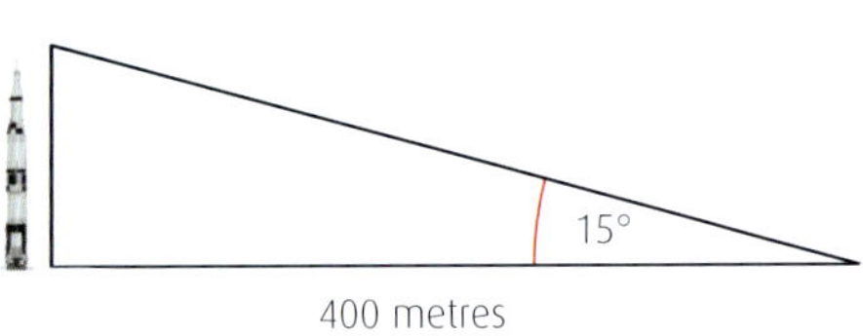

5 Use a scale of 1 cm to 40 m.
Calculate the height of this cliff.

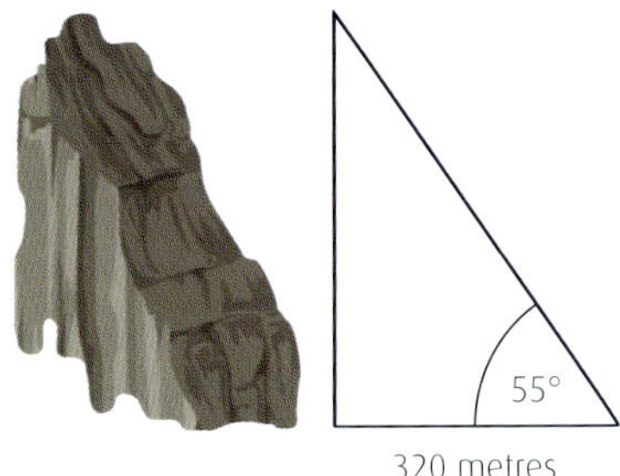

6 Use a scale of 1 cm to 400 m.
How high is the hang glider flying?

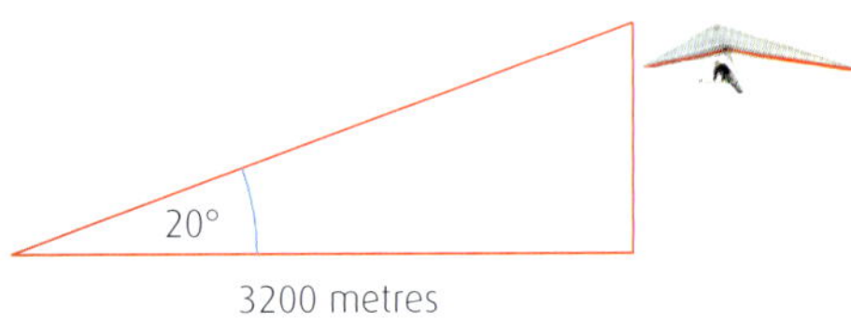

COORDINATES

I can use my knowledge of the coordinate system to plot and describe the location of a point on a grid. MTH 2-18a/MTH 3-18a

What's coming up?

This Outcome and Experience will give you the opportunity to:
- extend Level 2 coordinate work.

What you already know

You have already learned how to:
- ✔ describe, plot and record the location of a point, in the first quadrant, using coordinate notation.

How does that work?

You will recognise the coordinate grid for the first quadrant.

To plot the point **A (2, 4)**:

- Start at the origin O (0, 0).
- Move along the x-axis two squares.
- Move up parallel to the y-axis, four squares.

To plot the point **B (5, 1)**:
- Start at the origin O (0, 0).
- Move along x-axis 5 squares.
- Move up parallel to y-axis, 1 square.

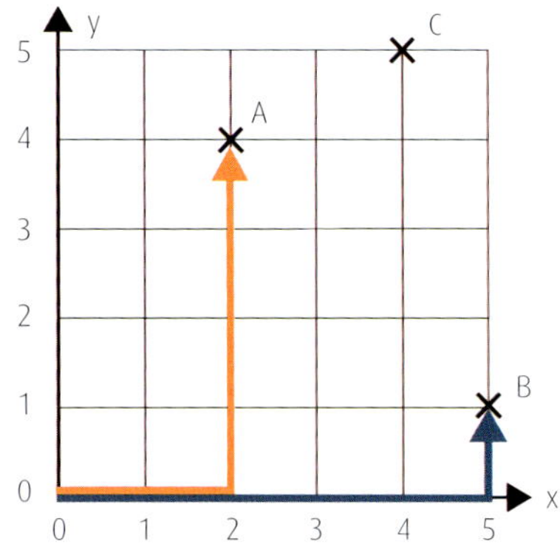

What are the coordinates of the point C?
- Start at the origin O (0, 0).
- Move along the x-axis until under point C : 4 squares.
- Move along y-axis until reach point C : 5 squares.

Reminder: All points are of form (x, y)
So point C has coordinates (4, 5)

DON'T FORGET

Remember, in the alphabet, x comes before y, so the x-coordinate comes before the y-coordinate!

Classroom challenge

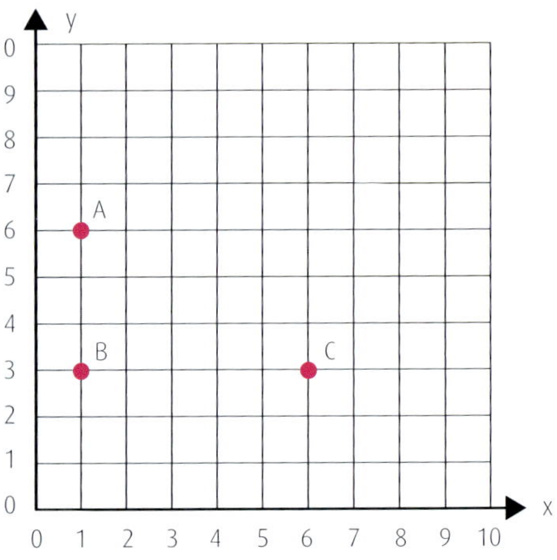

1 Make a copy of this coordinate grid.
  a Write down the coordinates of A, B and C.
  b Plot point D so that the four points make a rectangle.
  c Write down the coordinates of D.

2 On the same coordinate grid:
  a Plot the points; E (7, 6), F (7, 8), G (9, 8) and H (9, 6).
  b What shape have you drawn?

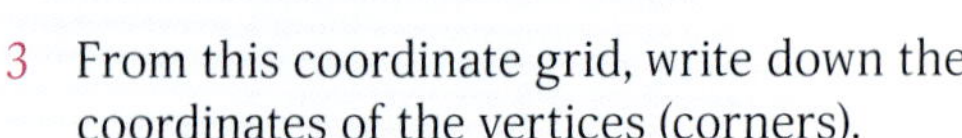

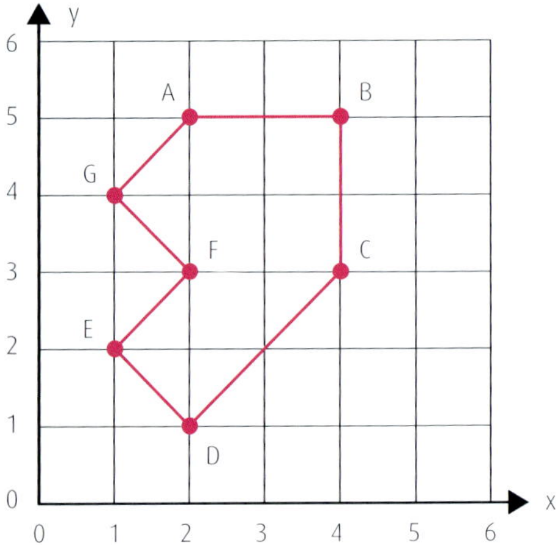

3 From this coordinate grid, write down the coordinates of the vertices (corners).

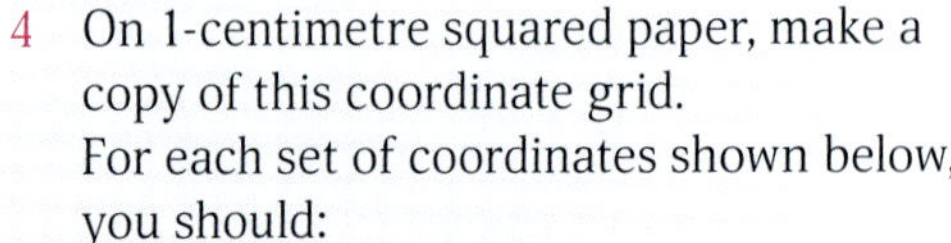

4 On 1-centimetre squared paper, make a copy of this coordinate grid.
For each set of coordinates shown below, you should:
  a plot the points
  b use a ruler to join the points in order
  c write down the name of the shape.

You may wish to use a different colour for each shape.

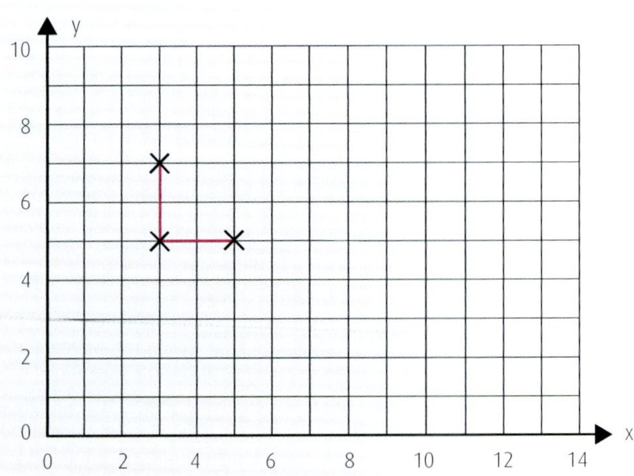

Shape A has been started for you.

Shape A	Shape B	Shape C	Shape D	Shape E	Shape F	Shape G	Shape H
(3, 7)	(7, 4)	(11, 0)	(0, 8)	(8, 10)	(7, 0)	(2, 2)	(12, 7)
(3, 5)	(9, 5)	(12, 2)	(0, 10)	(8, 7)	(5, 0)	(0, 5)	(13, 7)
(5, 5)	(11, 4)	(14, 2)	(3, 10)	(9, 7)	(4, 1)	(0, 0)	(14, 8)
(5, 7)	(9, 3)	(13, 0)	(0, 8)	(9, 10)	(6, 2)	(2, 1)	(14, 9)
(3, 7)	(7, 4)	(11, 0)		(8, 10)	(8, 1)	(2, 2)	(13, 10)
					(7, 0)		(12, 10)
							(11, 9)
							(11, 8)
							(12, 7)

5 Copy and complete the table to state the coordinates of each shape on the grid.

Shape	Coordinates
Triangle	
Cross	
Heart	
Square	
Lightning	
Circle	
Diamond	
Star	
Music Note	
Moon	

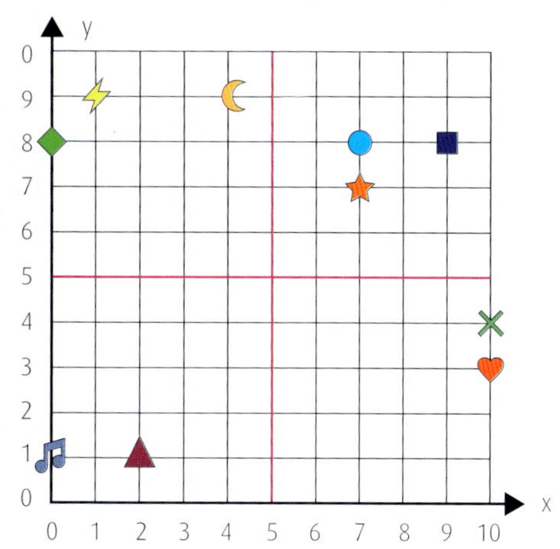

6 Make a copy of this grid.
  a Plot the points A (2, 1), B (2, 4) and C (5, 1).
  b Join the points to form a triangle.
  c Keep point A in the same place. Draw an enlargement of the triangle with scale factor 2.
  d Write down the coordinates of the new points B and C.

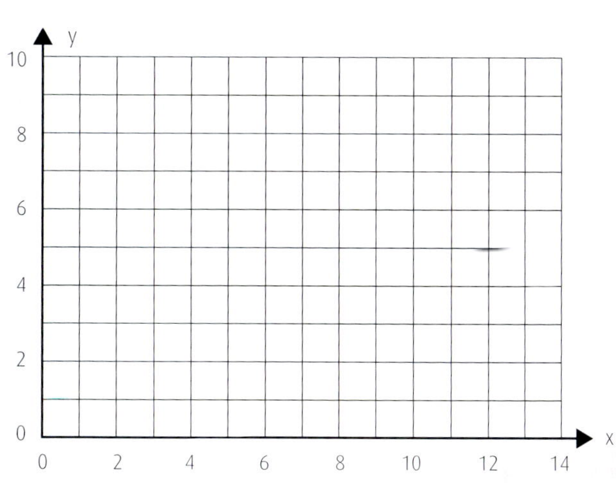

7 Draw axes for x and y from 0 to 10 on 1-centimetre squared paper.
  a Plot the points, A (2, 0), B (7, 0), C (9, 6) and D (5, 6). Join the points in order.
  b What is the name of this shape? What is its area?
  c Plot the points, E (3, 1), F (9, 1) and G (6, 9). Join the points in order.

d What type of triangle is this? What is its area?
e Plot the points, H (2, 4), I (2, 6), J (7, 8) and K (7, 2).
f What is the name of this shape? What is its area?
g Plot the points, L (0, 4), M (3, 6), N (6, 4) and P (3, 2). Join the points in order.
h What is the name of this shape? What is its area?

8 Use the grid below to solve this puzzle.
 a Copy the table and insert the letters that correspond to the coordinates. Show your teacher!

Coordinate	(5,7)	(9, 2)	(0, 8)	(1, 3)	(5, 7)	(2, 0)	(5, 7)	(9, 5)	(4, 9)	(1, 9)	(3, 4)
Letter											

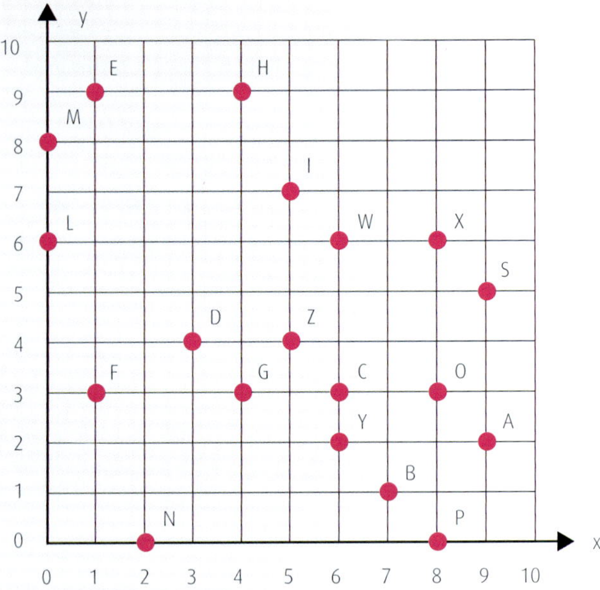

 b Write down the coordinates to spell the word LEMON.
 c Write down the coordinates to spell the word MEAN.
 d Write down the coordinates to spell the word MEDIAN.
 e Write down the coordinates to spell the word MODE.
 f Write down the coordinates to spell the word XENON.
 g Write down the coordinates to spell the word ZODIAC.

SYMMETRY

I can illustrate the lines of symmetry for a range of 2D shapes and apply my understanding to create and complete symmetrical pictures and patterns.
MTH 2-19a/MTH 3-19a

What's coming up?

This Outcome and Experience will give you the opportunity to:
- identify all lines of symmetry in 2D shapes
- create symmetrical patterns and pictures.

What you already know

You have already learned how to:
- ✓ identify and illustrate line symmetry on a wide range of 2D shapes and apply this understanding to complete a range of symmetrical patterns, with and without the use of digital technologies.

Symmetry is when one shape becomes **exactly** like another when you 'flip', 'slide' or 'turn' it.

The types of symmetry you will see are:

Bilateral symmetry	(you may see this called **line**, **mirror** or **reflection** symmetry)
Rotational symmetry	(you may see this called **turning** symmetry)
Translation symmetry	(you may see this called **sliding** symmetry)

Bilateral symmetry

How does that work?

Bilateral symmetry is when one 'half' of a shape will fold **exactly** on top of the other half.

The 'fold line' is called the **line of symmetry**.

In this picture, one half of the triangle would fold exactly on top of the other half.
This shape has **one line (or axis) of symmetry**.

This regular pentagon has **5 lines of symmetry**.

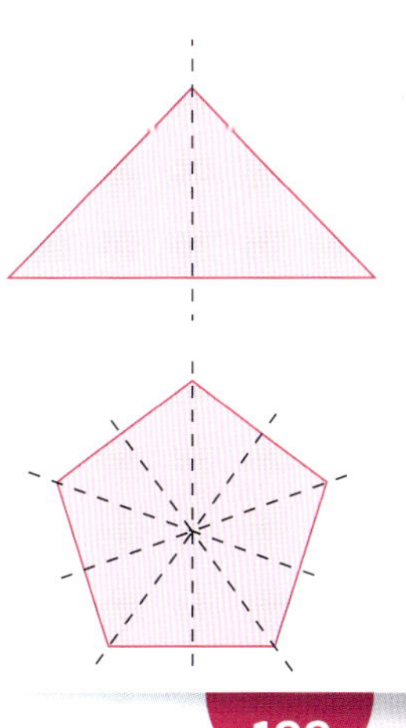

EXAMPLE

Complete this shape to make it symmetrical about the dotted line.

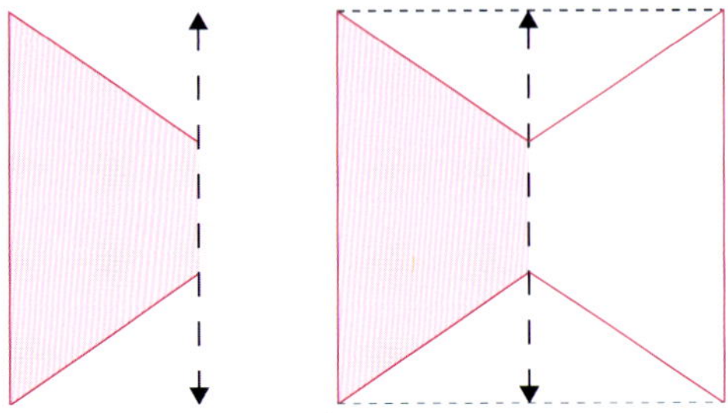

Classroom challenge

1 For each of these shapes, say whether the dotted line is a line of symmetry. For each shape write 'yes', or 'no'.

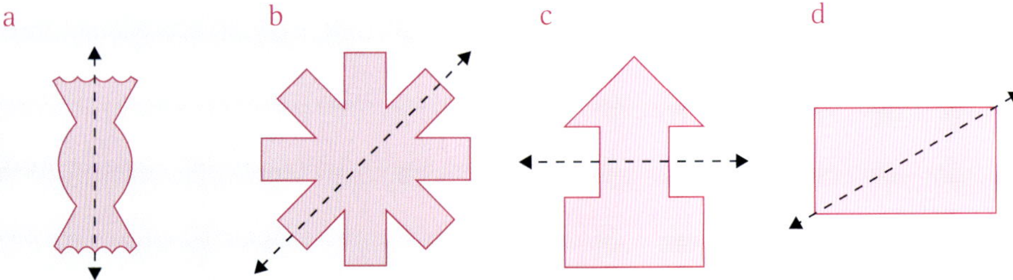

2 Copy, or trace, these shapes and draw in one line of symmetry.

3 Copy, or trace, these shapes. Mark in as many lines of symmetry on each shape as you can.

a

b

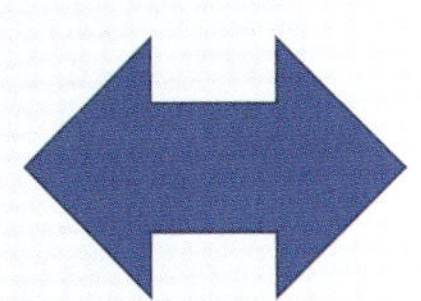

c

d

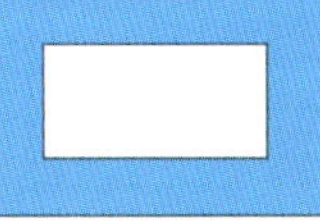

e

f

g

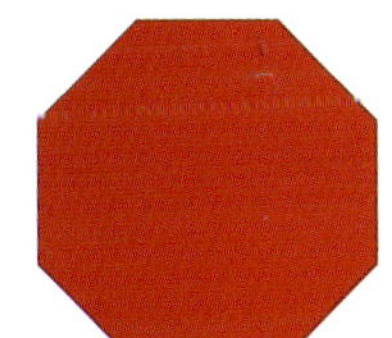

h

4 Copy or trace these shapes and then complete the other half to make a symmetrical shape.

a

b

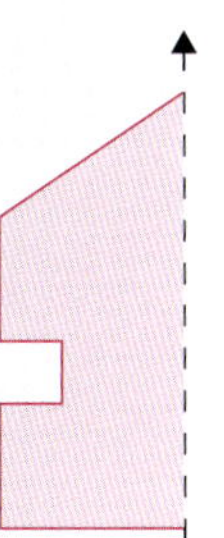

c

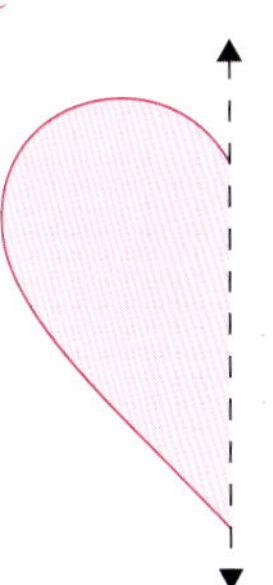

d

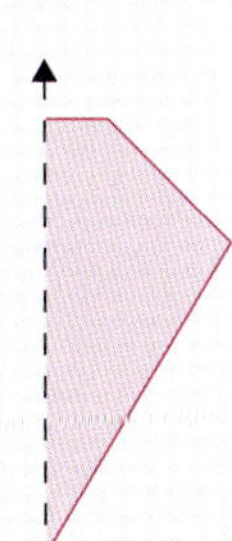

e

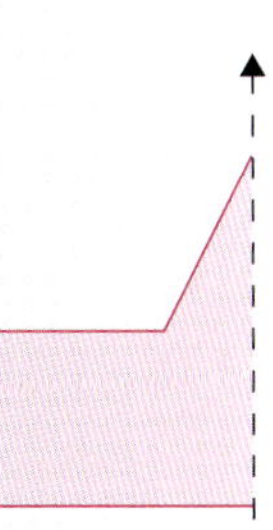

f

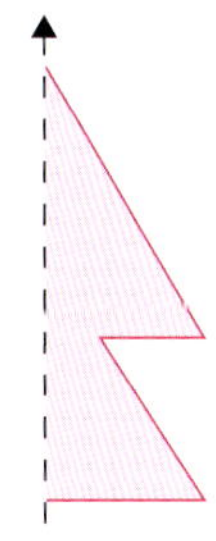

5 Draw these shapes on 1-centimetre squared paper.
Complete the 'reflection' to make a symmetrical shape.

a

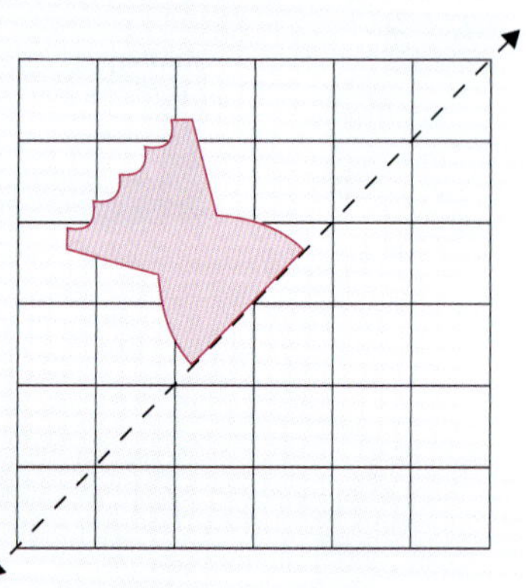

c

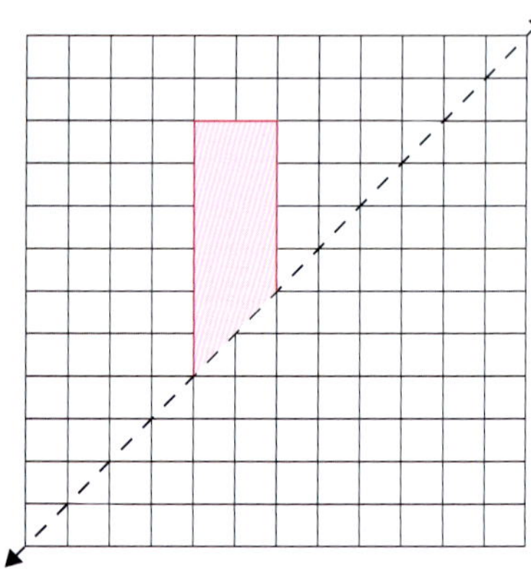

b

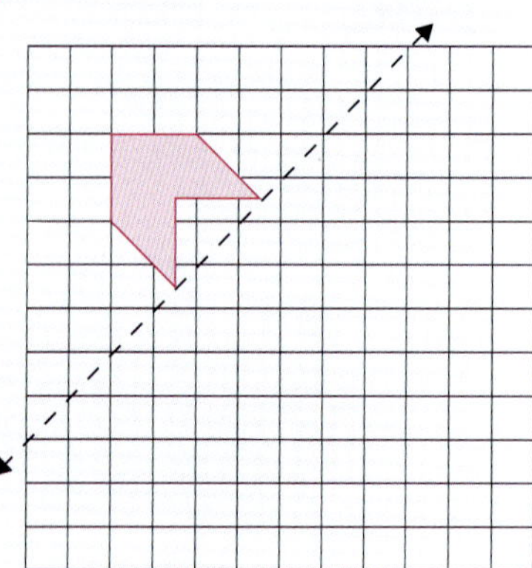

d

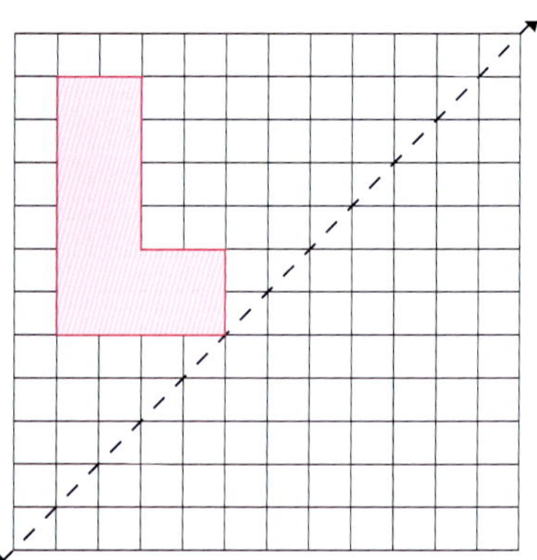

INFORMATION HANDLING

Mathematics is important in our everyday life, allowing us to make sense of the world around us and to manage our lives. Using mathematics enables us to model real-life situations and make connections and informed predictions. It equips us with the skills we need to interpret and analyse information, simplify and solve problems, assess risk and make informed decisions.

DATA HANDLING SKILLS

DATA AND ANALYSIS

I can work collaboratively, making appropriate use of technology, to source information presented in a range of ways, interpret what it conveys and discuss whether I believe the information to be robust, vague or misleading. MNU 3-20a

When analysing information or collecting data of my own, I can use my understanding of how bias may arise and how sample size can affect precision, to ensure that the data allows for fair conclusions to be drawn. MTH 3-20b

I can display data in a clear way using a suitable scale, by choosing appropriately from an extended range of tables, charts, diagrams and graphs, making effective use of technology. MTH 2-21a/MTH 3-21a

What's coming up?

This Outcome and Experience will give you the opportunity to:

- source information or collect data making use of digital technology where appropriate
- interpret data sourced or given
- describe trends in data using appropriate language, for example increasing trend
- determine whether information is robust, vague or misleading by considering, for example, the validity of the source, scale used, sample size, method of presentation and appropriateness of how the sample was selected
- collect data by choosing a representative sample to avoid bias
- organise and display data appropriately in a variety of forms, for example compound bar, line graphs and pie charts, making effective use of technology as appropriate.

What you already know

You have already learned how to:

- ✔ devise ways of collecting data in the most suitable way for the given task
- ✔ collect, organise and display data accurately in a variety of ways including through the use of digital technologies, for example by creating surveys, tables, bar graphs, line graphs, frequency tables, simple pie charts and spreadsheets
- ✔ analyse, interpret and draw conclusions from a variety of types of data
- ✔ draw conclusions about the reliability of data, taking into account, for example, the author, the audience, the scale and sample size used
- ✔ display data appropriately making effective use of technology and choose a suitable scale when creating graphs.

Collect, interpret and describe trends in data

What is data and information?

	Data	Information
Meaning	Data is the raw, or unorganised, facts or numbers that need to be processed. Data may seem simple, random, or useless – until it is organised.	When data is processed, for example, by organising it, analysing it or presenting it in a graph or chart, so as to make it useful, then it becomes information.
Example	The age of each person attending a concert is one piece of data.	However, if you calculate the average age of people attending the concert or put the ages into a chart showing 'age bands' then the data now presents information which is useful.
Origin of word	Data is plural. The singular is datum. It originates from Latin and means 'something given'.	Possibly originates from Old French or Middle English. It means 'communication of news' or 'knowledge communicated relating to a particular topic'.

Working with charts and tables

How does that work?

Julie's Just Giving is a gift shop that sells gifts for all occasions. She records her sales over the last four years.

Gift Occasion	2016	2017	2018	2019
Birthday	90	90	101	120
Wedding	65	72	58	50
Anniversary	72	83	62	58
Retirement	35	26	37	36

a How many gifts did Julie sell in 2018?
b How many retirement gifts were sold over the four years?
c What is the trend in sales of birthday gifts?

SOLUTION

a Julie sold 101 + 58 + 62 + 37 = 258 gifts in 2018.
b Retirement gifts 35 + 26 + 37 + 36 = 134 over the four years.
c The sales of birthday gifts are generally increasing over the four years.

Classroom challenge

1 The chart shows the scores of five students in a science class.

	Term 1	Term 2	Term 3
Ajay	62	71	73
Becky	43	65	57
Choi	71	75	82
David	63	37	58
Erica	82	73	64

a What was Becky's total score over the three terms?
b Who scored the highest in Term 2?
c What was the lowest score in Term 3?
d Describe the trend in scores for (i) Ajay and (ii) Erica.

2 Choux Bakery kept track of the pastries it sold over one week.
The results are shown in the table.

Flavour	Monday	Tuesday	Wednesday	Thursday	Friday
Cream	60	70	20	65	52
Strawberry	51	43	17	61	52
Chocolate	58	57	30	60	43

a What was the total number of pastries sold on Monday?
b How many more strawberry pastries were sold on Thursday than Tuesday?
c Choux Bakery closes for a 'half-day' once in the week. Which day do you think this is? Give a reason for your choice.
d Which was the most popular flavour over the week?

3 The chart shows the stopping distances for cars when braking in normal conditions.

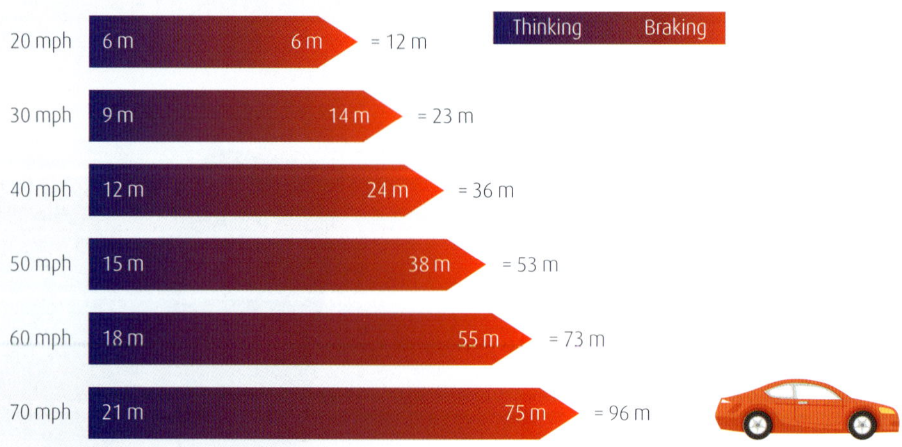

a What is the thinking distance for a car travelling at 50 mph?
b How much longer is the braking distance compared with the thinking distance at 30 mph?
c How much further would a car travel before stopping if it was going at 30 mph rather than 20 mph?
d The maximum speed limit on a motorway is 70 mph.
How many metres would it take a car to stop if travelling at this speed?
e John is travelling at 40 mph. A car pulls out from a side road 40 metres away. Should John be able to stop in time?

4 The table shows the men's long jump medallists at the Commonwealth Games held in the Gold Coast, Australia, in 2018.

Medal	Name	Country	Jump					
			1	2	3	4	5	6
Gold	Manyonga	South Africa	8·24	8·21	X	8·35	X	**8·41**
Silver	Frayne	Australia	8·00	**8·33**	X	8·08	X	X
Bronze	Samaai	South Africa	8·06	**8·22**	8·17	7·89	8·18	8·08

a What was the length of jump that won the gold medal?
b In which jump were both the silver and bronze medals decided?
c How much further did Samaai jump in jump 6 compared with jump 4?
d What was the difference in length between the jumps for the gold and bronze medal positions?

5 The table shows the cost of a short break in the city of Bremen, Germany.

Dates	Number of nights	Cost in £	Number of nights	Cost in £	Cost for extra night(s)
Jan–Apr	2	312	3	367	50
May–Aug	2	357	3	412	51
Sept–Dec	2	371	3	426	53

From the table, find the cost of:
a a two-night break in June
b a three-night break in October
c a four-night break in January
d a five-night break in September.
e Selina has £370. Which breaks could she afford?

DON'T FORGET

Find cost of 3 nights and add 1 extra night

6 The table shows times and costs for flights leaving Edinburgh and flying to Marrakech.

Airline	Depart	Arrive	Cost in £
Jet-Ease	1100	2015	543
Siberia Air	0600	1725	556
Ronan Air	1015	2245	765
Wavy Air	1805	0705 (next day)	467

a What is the difference in price between Wavy Air and Siberia Air?
b How long is the Jet-Ease flight?
c What is the difference between the cheapest and most expensive fares?
d Why do you think the Wavy Air flight is cheaper than the others?

7 The table shows a magazine's special offer for subscriptions.

Number of months subscription	Full price	Special offer price
6	£19·50	£17·55
12	£39·00	£33·15
18	£58·50	£46·40
24	£78·00	£58·50

a How much is saved if the 6-month subscription is taken?
b What percentage is this of the full price?
c What does the magazine normally cost per month?
d How much is saved by buying 24 months' subscription rather than two 12-month subscriptions?

8 On his birthday each year, Grant measured his height and marked it on a chart.

a What height was Grant at age 11?
b How old was Grant when he reached 165 centimetres?
c Between which ages did Grant grow most?
d Describe the trend of the graph.

9 This chart shows the pet sales from 'Petter Care' pet shop over the first five months of 2018.

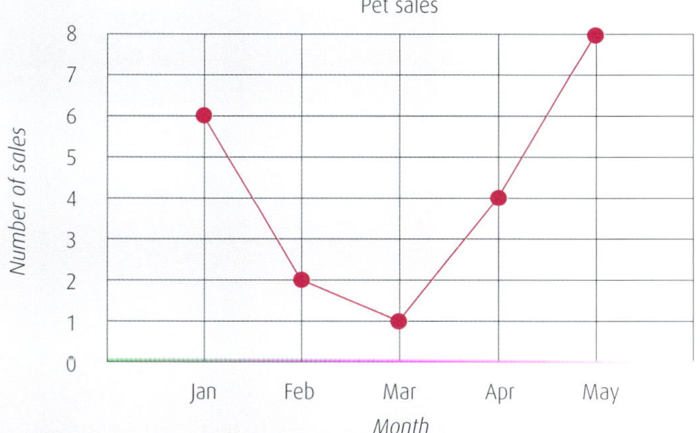

a How many pets were sold in April?
b How many pets were sold in May?
c Were more pets sold in January or April?
d In which month were most sales made?
e What was the trend in sales from January to March?

10 The chart shows the categories relating to your body mass index (BMI). The BMI is calculated by a formula which uses your height and weight.

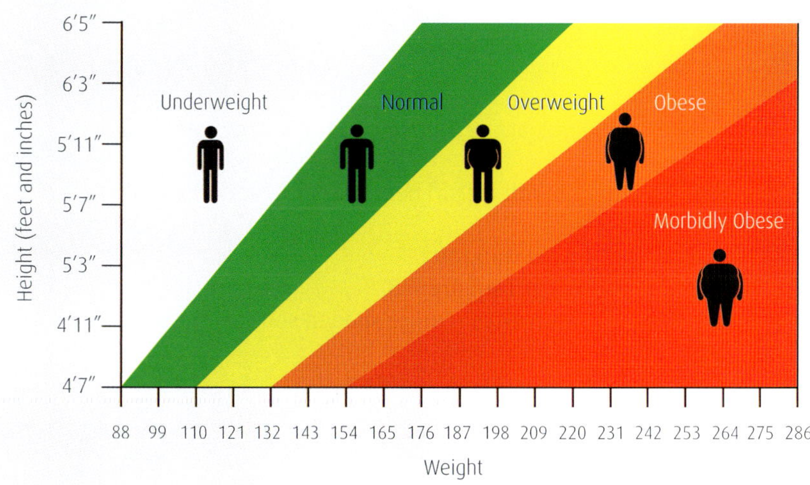

a Sam is 5' 11" and weighs 245 lbs. In which category does he fall?
b Erin is 5'7" and weighs 135 lbs. In which category does she fall?
c Colin is 6' 3" and 210 lbs. In which category is Colin?
d Graeme is 5' 7" and weighs 110 lbs. How many pounds would he need to gain to move from underweight to normal?

Working with graphs

Being able to work with different graphs will allow you to make informed decisions on the data you are looking at. Graphs come in many forms, some of which you have seen in the previous exercise.

Here, you will practise working with a wider range of graphs.

How does that work?

The **compound bar graph** shows the results of a survey into how many pets a group of children had.

Use the graph to answer these questions.

a How many girls do not have a pet?
b How many children own hamsters?
c Are hamsters more popular with girls or boys?
d How many boys have budgies?
e What is the most popular pet with boys?

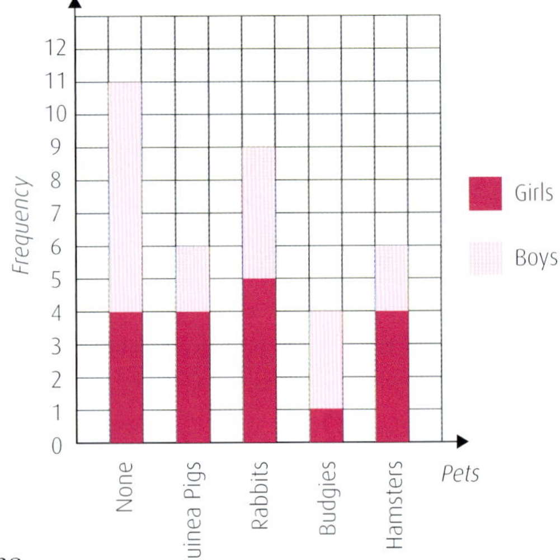

SOLUTION

a 4 girls do not have a pet. Look at the 'none' column. There are 4 squares like this:
b 6 children own hamster. Look at the hamster column – the total height is 6.
c Hamsters are more popular with girls. There are 4 'girl squares' and 2 'boy squares'.
d 3 boys have budgies: 3 shaded squares .
Or, height of column = 4 – 1 girl square = 3.
e Rabbits are most popular with boys – more shaded squares than other columns.

Classroom challenge

1 The bar chart in the example on page 212 could have been drawn like this.

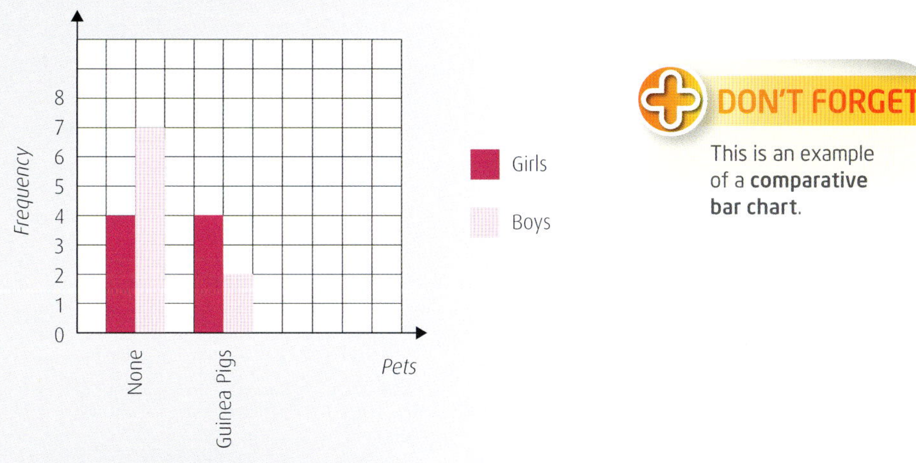

DON'T FORGET

This is an example of a **comparative bar chart**.

a Copy and complete the bar chart.
b Which bar chart do you think presents the data more clearly?

2 The table shows the results of a class survey on how the students travelled to school.

Method of travel	Tally	Frequency
Walk	~~||||~~ ||||	9
Bus	~~||||~~ ~~||||~~ ||	12
Bike	|||	3
Car	~~||||~~		6
	Total	30

a Illustrate this data using a bar graph.
b Copy and complete this table to help you to illustrate the information in a **pie chart**.

Note that there are 360° in a circle.
The calculation is used to find the size of the angle at the centre of the pie chart for each method of travel.
For example, 9 out of 30 students walked, giving:

$\frac{9}{30} \times 360 = 108°$

Method of travel	Frequency	Calculation	Angle
Walk	9	$\frac{9}{30} \times 360$	108
Bus	12	$\frac{}{30} \times 360$	
Bike	3	$\frac{}{30} \times$	
Car	6	$— \times$	
Total	30		360

3 The table shows data collected from a year group when they were asked to name their favourite colour.

Colour	Girls	Boys
Blue	7	8
Green	4	4
Red	8	5
Yellow	2	4
Orange	3	5
Purple	4	3
Pink	2	1

a Illustrate this data on a compound bar graph like the one in the example.
b Illustrate this data in a comparative bar chart like the one in question 1.
c Which was the most popular colour with the boys?
d Which was the most popular colour with the girls?
e Which colour was equally well-liked by both boys and girls?

4 The bar chart shows the direct sales, and online sales for four companies.
a Which company had the most **total sales**?
b Which company had fewer online sales than direct sales?
c For Company B, what percentage of its total sales was due to online sales?
d For Company D, what percentage of its total sales was due to direct sales?

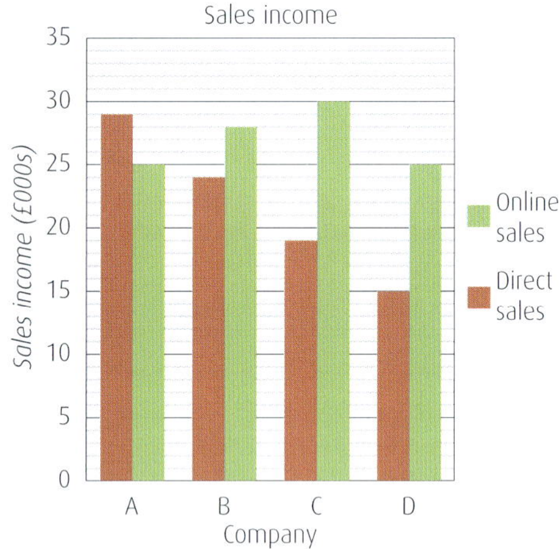

5 A company manufactures three different types of grass cutter: type A, type B and type C.
The production figures for a four week period are shown in the graph.

a In which week were most grass cutters produced?
b In week 4, how many more type B than type C were made?
c In week 3, what percentage of total production was type A?
d Over the 4 weeks, which type was produced the most?
e To save on costs, the company is thinking of stopping production of one type of grass cutter.
Which one do you think they should stop producing? Give a reason for your choice.

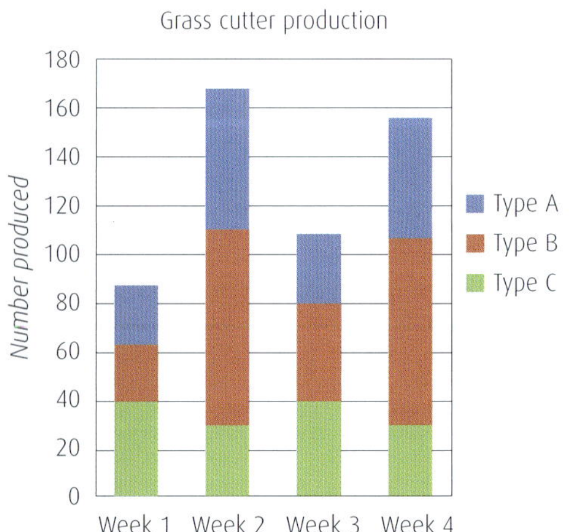

6 The graph shows information collected by a hotel on its guests during one week in June.

a How many guests, in total, stayed at the hotel?
b What fraction of the guests were Irish?
c How many guests, from the Rest of the World, were female?
d Approximately what percentage of the total guests staying were British males?

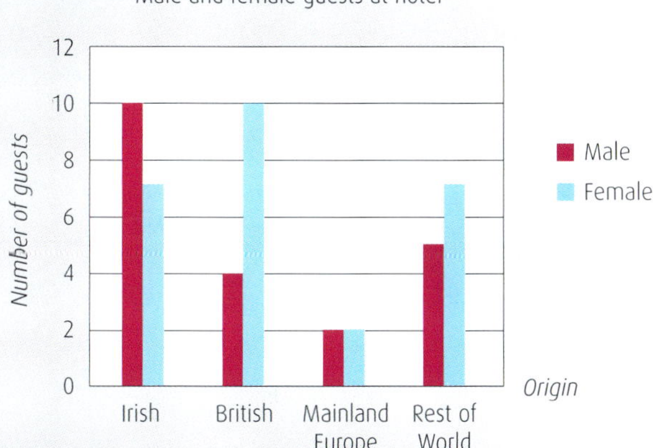

7 The hotel then further broke the data down to look at the ages of guests staying during that week. The results are in the graph below.

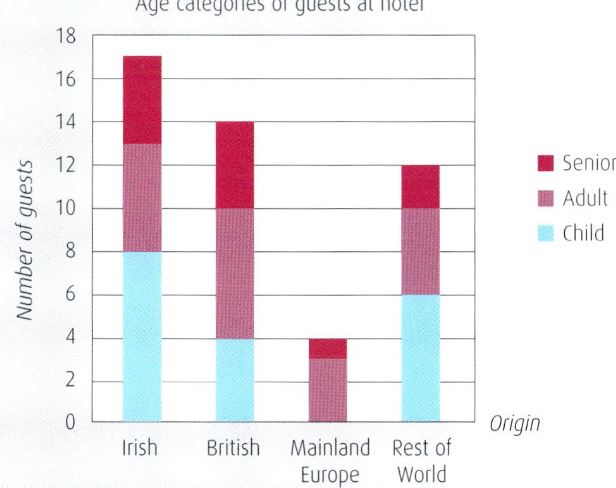

a How many of the guests were British children?
b How many of the guests were Irish adults?
c How many of the guests were Seniors from mainland Europe?
d How many children in total stayed at the hotel?
e What percentage of Rest of the World guests were adults?

8 This pie chart shows how employees in a call centre get to work each day.

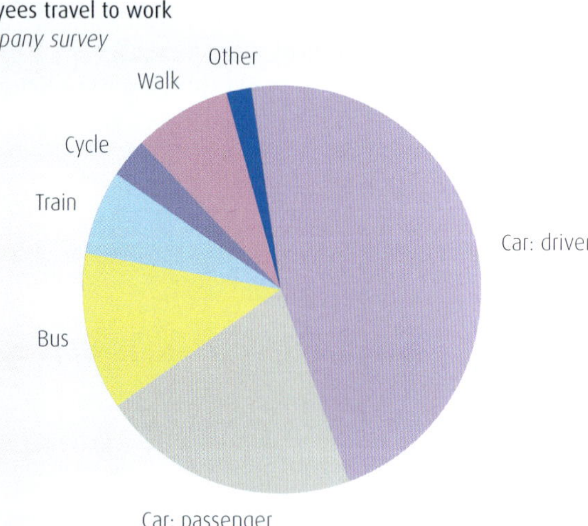

a Do more employees travel to work by bus or by train?

b Is it correct to say that there are about twice as many employees who travel to work as car drivers compared with those who travel to work as car passengers? Give a reason for your answer.

c Estimate the proportion of employees who get to work by methods other than by car.

Bias and sample size

How does that work?

Penny thinks that the dice she is using in her game of Monopoly is unfair because she never seems to get a six when she wants one.
Penny decides to try to check the dice and she rolls it 60 times.
The diagram shows the results of each roll.

a Illustrate the results in a diagram.

b Do you think Penny's dice is fair?

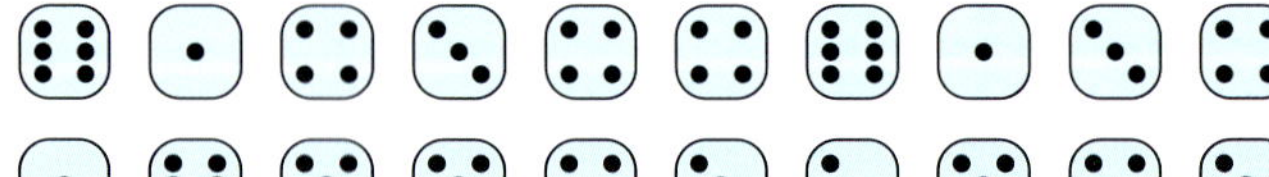

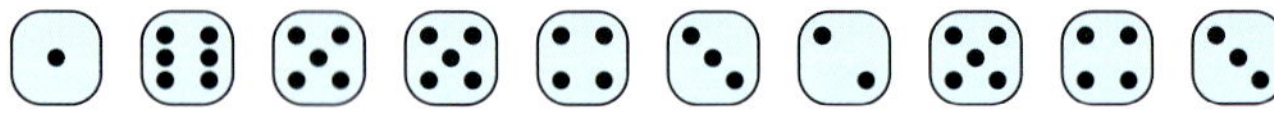

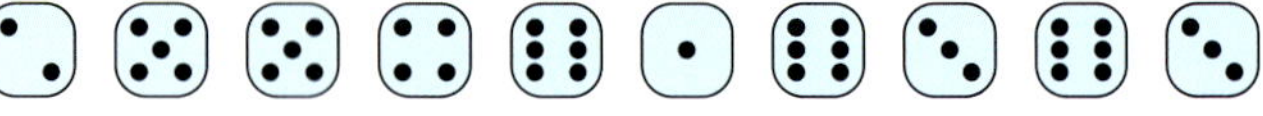

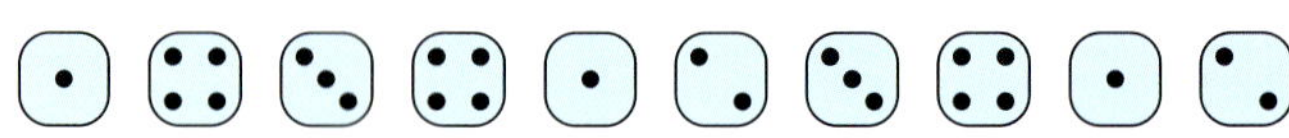

We have used the spelling dice instead of die throughout the book as more commonly used in speech.

SOLUTION

a

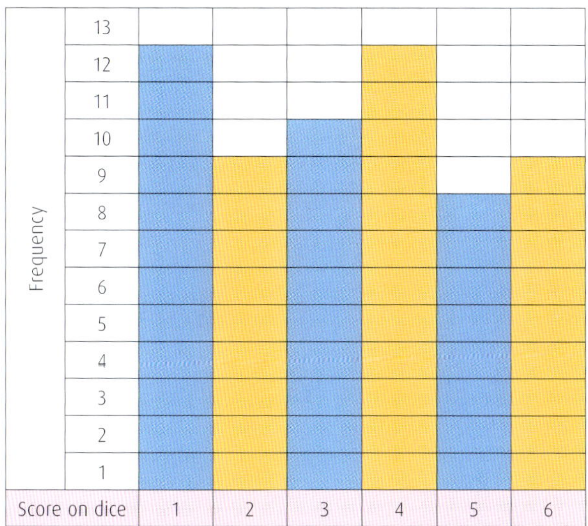

b The diagram seems to indicate that there may be a slight bias.
There seems to be a lot of 1s and 4s compared with other scores.
However, the sample size is small – if you rolled the dice many more times you would expect to get a six about $\frac{1}{6}$ of the time.
In this trial, that would be about 10 in every 60 rolls.
Penny got 9 sixes so that would seem fair.

Classroom challenge

1 The two graphs show the annual snowfall in a city in the years 2016 and 2017.
Which graph gives the impression that the annual snowfall in the city was twice as much in 2017 than 2016?
Explain why this graph is misleading.

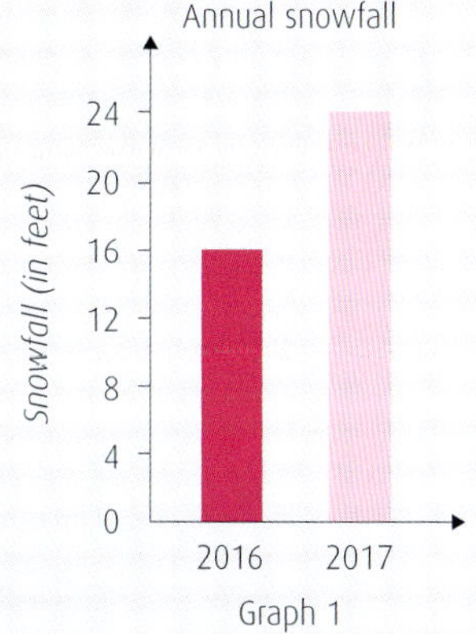

Graph 1

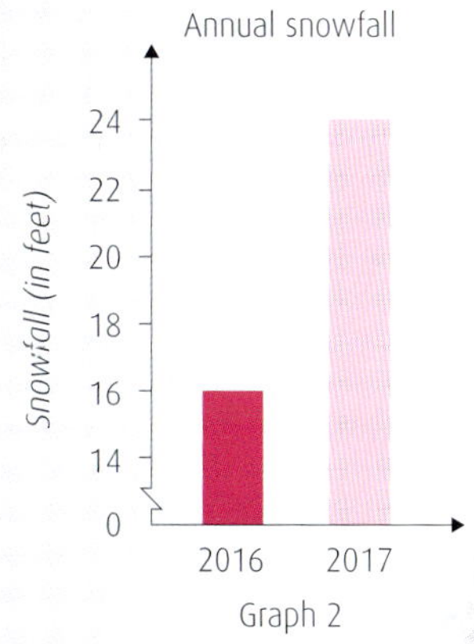

Graph 2

2 The graph shows the number of toys sold in a toy shop over the first five months of the year.

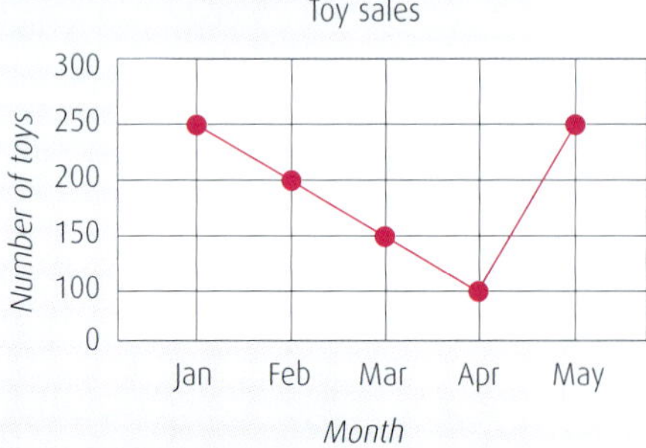

Explain why this graph is misleading.

3 The bar graph illustrates the life span of polar bears and whales.

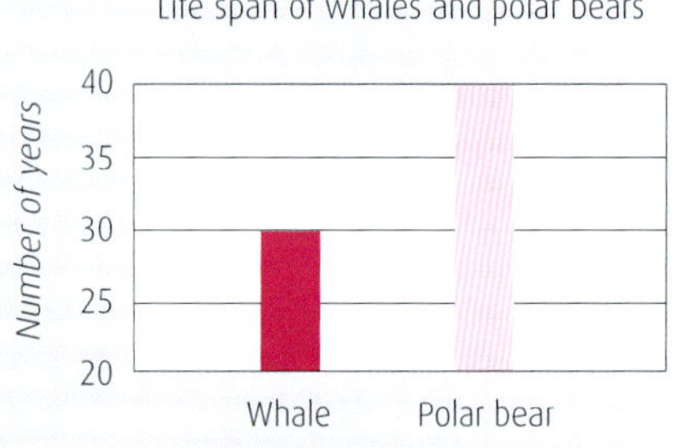

What change would you make to the graph so that it is not misleading?

4 The graph shows the number of endangered reptiles in Zimbabwe and Ghana.

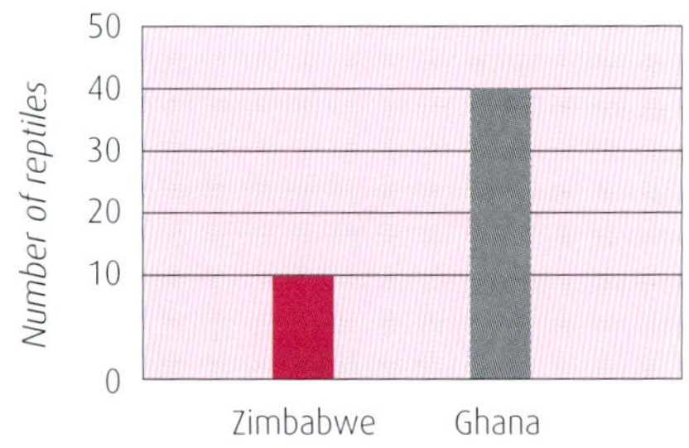

a Is the graph misleading?
b If it is, what would you do to make it not misleading?

5 The graphs show the weight-lifting capacity of three machines.

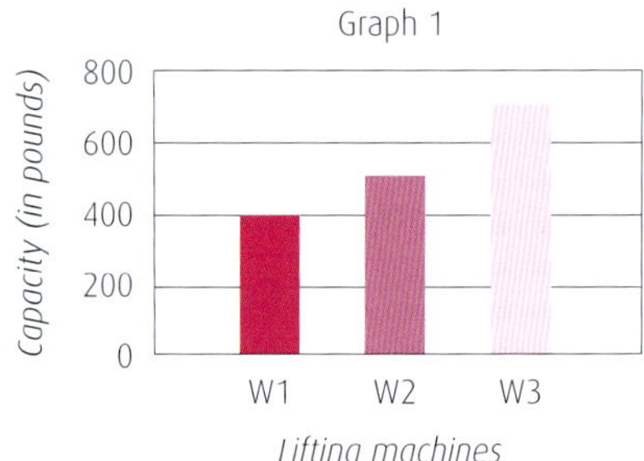

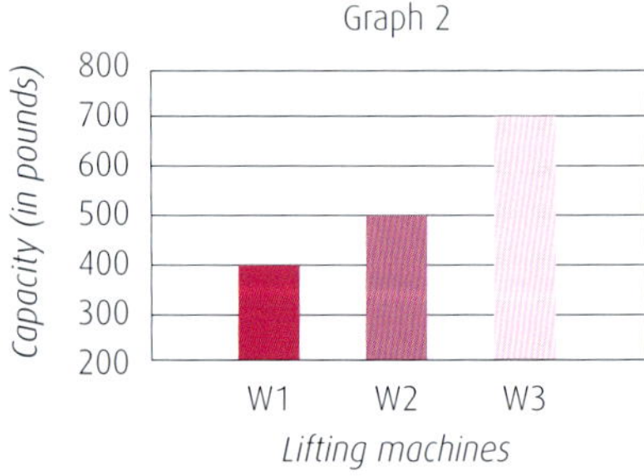

Which graph is misleading, and why?

6 Finlay is trying to impress his mum by showing her how his maths test marks have improved.

a Which graph should Finlay use?

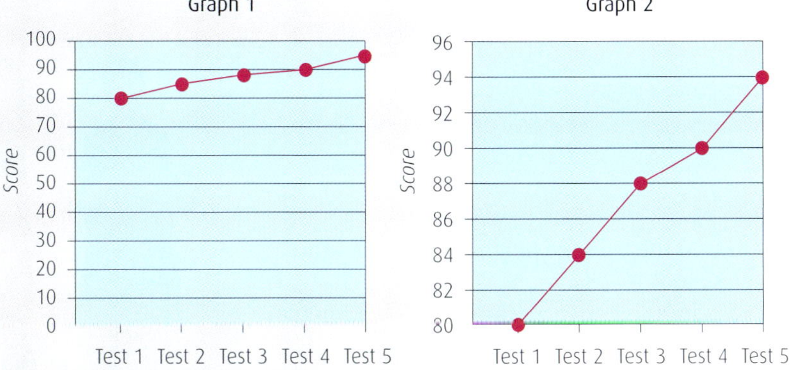

b Finlay's pal, Connor, says he is misleading his mum.
Is Connor correct? Explain your answer.

7 The graph illustrates the number of some of the species of birds in a zoo.
Explain why the graph is misleading.

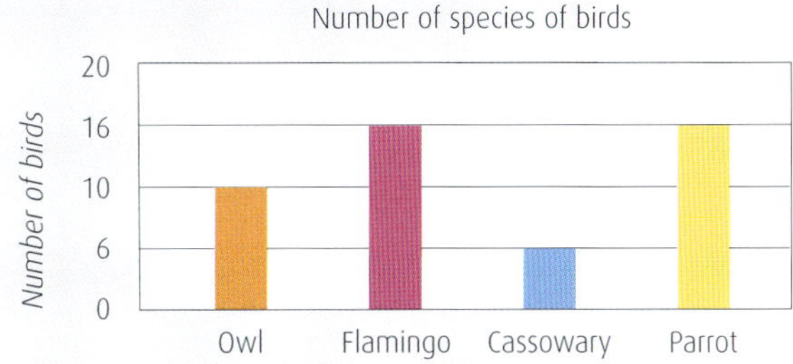

8 Cohen is using this spinner to play a game.
He thinks the spinner may be biased.
He spins it 50 times and records the results.

Lands on number	Frequency
1	11
2	9
3	7
4	13
5	10

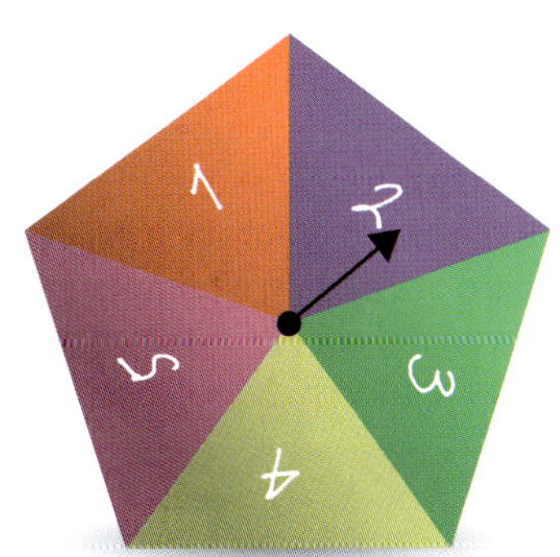

Do you think the spinner is fair? Explain your answer.

9 In each of the examples below, say whether the sample is likely to be biased or unbiased.

	Question	Sample is chosen from
a	What is your favourite sport?	People attending a cricket match
b	What is your favourite fruit?	People listed in a phone book
c	Should more money be put into school athletics or school music programmes?	People attending band rehearsals
d	What is your favourite holiday destination?	All students in the class

STRETCH YOURSELF

10 You want to find the average height of students who attend your school.

a Describe a method of collecting data, for a sample, which will give a biased result.

b Describe a method of collecting data, for a sample, which will give an unbiased result.

Organise and display data

In this section, you will construct and interpret information given in a variety of ways.

Different types of diagrams and graphs are useful for displaying information.

Here, we will look at pie charts, line graphs and stem-and-leaf diagrams.

Pie charts

How does that work?

EXAMPLE

At the end of term, in Westwade High School, students are graded A, B, C or D for their reports.
The table (right) shows the information for class 1M2.
First, the angle for each grade needs to be calculated.
There are 360° in a complete circle.
As you did earlier in the section, add another column to the table.

Grade	Frequency
A	11
B	10
C	7
D	2
Total	30

Grade	Frequency	Calculation	Angle
A	11	$\frac{11}{30} \times 360$	132°
B	10	$\frac{10}{30} \times 360$	120°
C	7	$\frac{7}{30} \times 360$	84°
D	2	$\frac{2}{30} \times 360$	24°
	30		360°

Now you need to draw the blank pie chart and start to mark in the sectors. Both the table and the pie chart illustrate the data clearly.
It is easy to see that few students got a poor grade (grade D), and that the most common grade was a grade A.

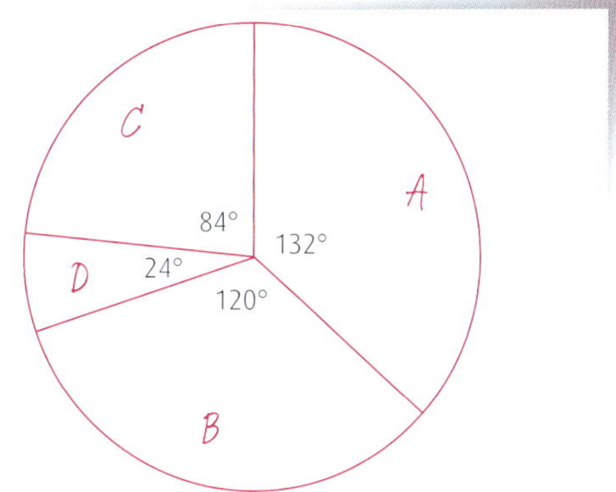

Classroom challenge

1 A group of 40 students were asked to state their favourite colour.
The results are in the table below.
Copy the table and complete the last two columns.
Construct, and label, a pie chart to illustrate this data.

Colour	Number of students	Calculation	Angle
Blue	7	$\frac{7}{40} \times 360$	63°
Red	20		
Yellow	8		
Other	5		
	40		360°

For each question in this exercise, you can either draw a 5 centimetre circle to use as a pie chart or download the pie chart template in our Homework Helpers.

2 A class of 20 students were asked what lesson they had last period on a Friday.
The results are shown in the table below.
Copy the table and complete the last two columns.
Construct, and label, a pie chart to illustrate this data.

Subject	Number of students	Calculation	Angle
Mathematics	8	$\frac{8}{20} \times 360$	144°
English	6		
French	2		
Art	4		
	20		

3 A group of 30 people were asked what type of pet they had.
The results are shown in the table below.
Construct, and label, a pie chart to illustrate this data.

Type of pet	Dog	Cat	Gerbil	Snake	Other	Total
Number of people	12	7	6	1	4	30
Angle				12°		

4 The colour of cars in a car park one Saturday morning was noted.
The results are shown in the table.
Copy and complete the table.
Construct, and label, a pie chart to illustrate this data.

Colour	Frequency	Angle
White	15	
Red	11	
Blue	12	
Black	17	
Silver	5	
Total		360°

5 An optician noted the colour of eyes of people coming in to have an eye test.
The results are shown in the table.
Use this information to construct and label a suitable pie chart.

Eye colour	Blue	Green	Hazel	Grey	Brown	Other
Frequency	40	22	18	15	18	7

6 A hill-farmer took a note of how many of each animal she had on her farm.
Use the information given to construct a pie chart.

Type of animal	Cow	Goat	Pig	Sheep	Chicken
Number	90	12	25	34	19

7 Teigan is a car sales person. She kept a note of the makes of car she sold over a six-month period.
Her results are shown in the table.
Construct and label a pie chart to show this information.

Make of car	Ford	VW	Suzuki	Fiat	Renault
Frequency	12	18	18	2	10

8 Mr Khan went to his local store to do some shopping.
The pie chart illustrates the money he spent on various items.

a What did Mr Khan spend most money on?

b What fraction of money was spent on petrol?

c If Mr Khan spent £30 on petrol, how much did he spend in total at the store?

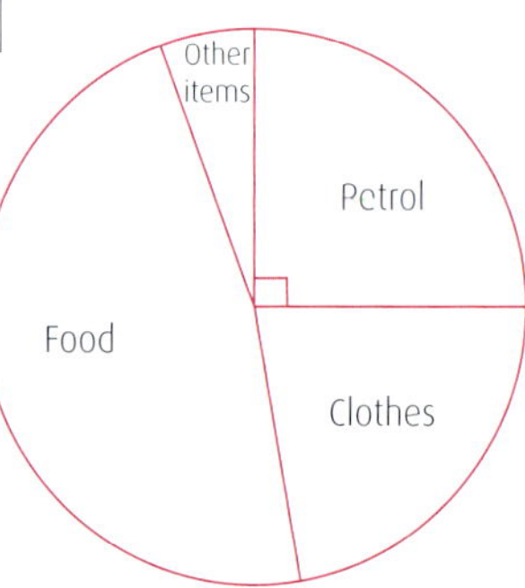

9 Ms Kirkpatrick asked her class of 24 students how they travelled to school that morning. The results are shown in the pie chart.

a How did most of her student travel to school that morning?

b How many students cycled to school?

c How many students walked to school?

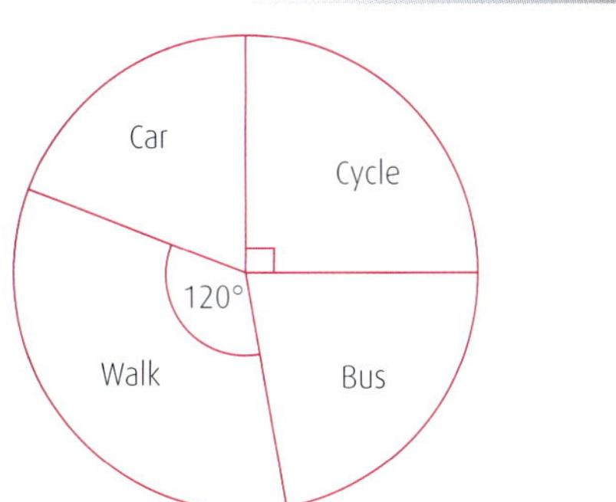

STRETCH YOURSELF

10 The pie charts show some information about medals won by Jamaica and Ghana at a recent Commonwealth Games.

Medals won by Jamaica

Gold
72°
168°
Bronze
120°
Silver

Medals won by Ghana

Jamaica won 7 bronze medals.

a How many medals, in total, did Jamaica win?

b How many gold medals did they win?

c Usain says 'The pie charts show that Jamaica won more gold medals than Ghana'. Is this statement correct? You should explain your answer.

Line graphs

Classroom challenge

1 The line graph shows the population of rhinos at a zoo over a six-year period.

a In 2013, what was the population of rhinos?

b In which year did the population reach 30 rhinos?

c Between which two years was the greatest increase in population?

d What was the increase in population from 2017 to 2018?

e What is the trend in population of rhinos?

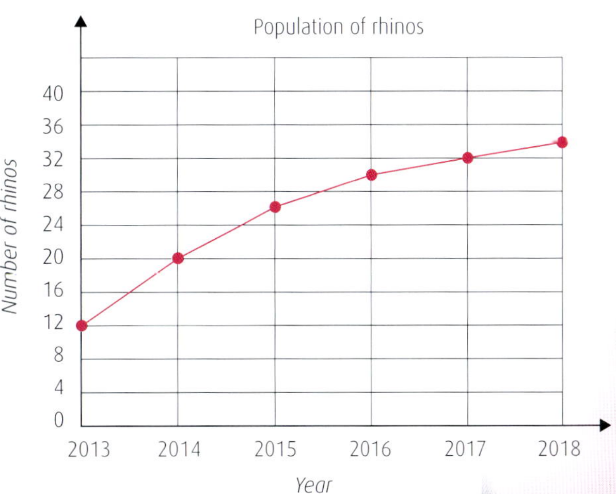

2 Charlie is a keen amateur weather-watcher.
He kept a record, over a year, of the average air temperature outside his house.
The results are shown in the graph.

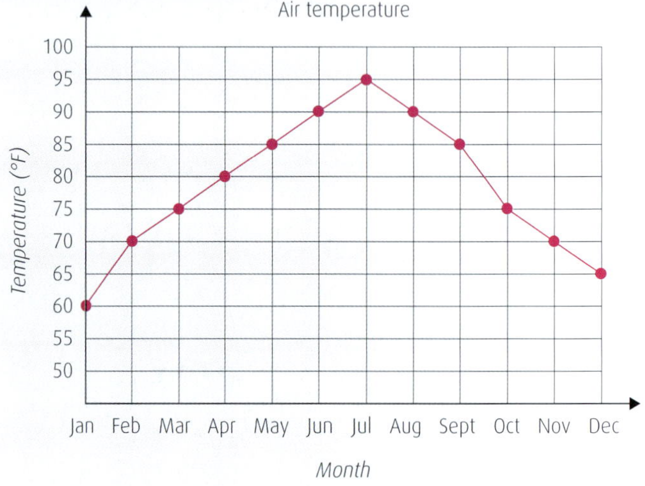

a What was the lowest temperature recorded?
b What was the highest temperature recorded?
c In which month(s) was a temperature of 85°F recorded?
d What is the trend in the graph for the last five months in the year?
e How much warmer was it in September compared with March?

3 George and Penny are doing a motorbike tour around Peru.
The number of miles they rode each day in the first week is given in the table below.

Day	Number of miles
Monday	40
Tuesday	70
Wednesday	55
Thursday	80
Friday	65
Saturday	50
Sunday	0

Copy the blank graph on page 225.
Plot the daily information and join up to construct a line graph.

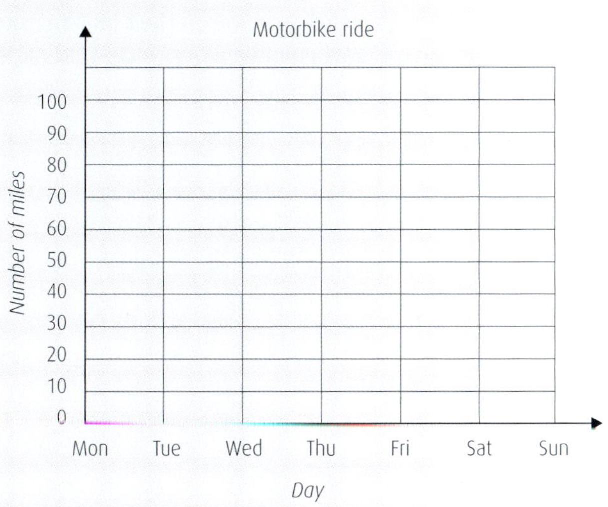

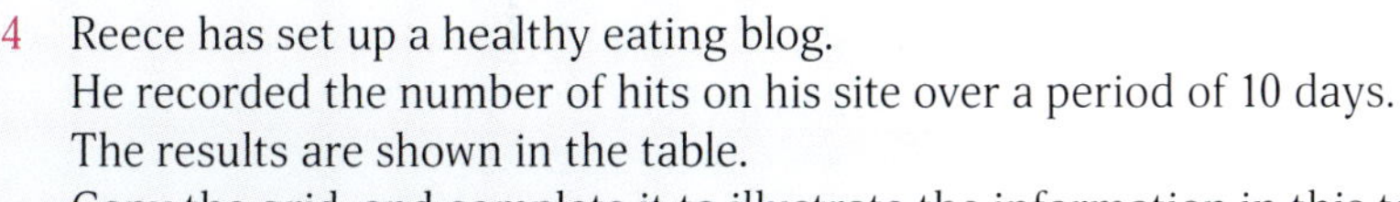

4 Reece has set up a healthy eating blog.
He recorded the number of hits on his site over a period of 10 days.
The results are shown in the table.
Copy the grid, and complete it to illustrate the information in this table.

Day	Number of hits
1	150
2	350
3	550
4	700
5	800
6	750
7	600
8	850
9	300
10	200

Healthy eating blog

Number of hits: 0, 100, 200, 300, 400, 500, 600, 700, 800, 900, 1000

Day: 1, 2, 3, 4, 5, 6, 7, 8, 9, 10

5 The number of students absent from first year classes in one week are shown in the table below.
Construct a suitable line graph to illustrate this information.

Day	Monday	Tuesday	Wednesday	Thursday	Friday
No. of students absent	21	14	11	24	35

6 Kala Building firm has a number of male and female staff working on their building sites. The graph shows the number of male and female staff employed over a six-year period.

a In which year were the most female staff employed?
b How many more male than female staff were there in 2015?
c When was the gap between number of male and female staff the smallest?
d Which years had a total number of 450 staff?
e In which year did the number of male staff fall but female staff increase?

7 The number of cars produced at two factories were recorded over a number of years.
The results are shown in the table on the right.
Choose suitable scales and construct a double line graph to illustrate this information.

Year	Carter Cars	Clinton Cars
2013	800	650
2014	950	750
2015	900	850
2016	1000	900
2017	1200	1000
2018	700	900

Stem-and-leaf diagrams

Stem-and-leaf diagrams are an excellent way of displaying data.

How does that work?

When constructing a stem-and-leaf diagram, which is like a bar graph, it is best to do it in three steps.

Step 1 Get the data on to the diagram.
Step 2 Order the data in the 'leaves'.
Step 3 Ensure that there is a 'key'.

EXAMPLE

The age in years, of 15 students on a college course are:

19	18	20	25	37
33	21	17	29	20
42	18	23	37	22

Draw a stem-and-leaf diagram to illustrate this data.

SOLUTION

Step 1 Set up the stem-and-leaf diagram.
This consists of the stem the 'first part' of the number
the leaves the final part of the number
Step 2 Order the leaves
Step 3 State the key

Step 1 The numbers are from the '10s' to the '40s' so the stem will be
The diagram starts like this:

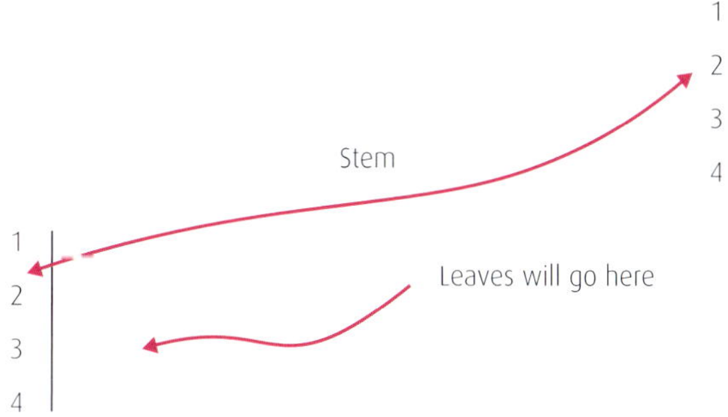

SOLUTION CONT

Now put in the leaves. It is easier to go along rows or down columns and enter numbers as you meet them – they can be put in order later.

1	9 8 7 8
2	0 5 1 9 0 3 2
3	7 3 7
4	2

These are from the top row of the table of data.
The second row is entered in red.
The third row is entered in green.

This way, we can be reasonably sure that we have entered all the data.
Step 2 Order the leaves.
Step 3 Insert a 'key'.

Age of college students

1	7 8 8 9
2	0 0 1 2 3 5 9
3	3 7 7
4	2

Key: 3|7 = 37 years
N = 15

N = Number of pieces of data

This data is now in a form in which we can look at the information it provides. For example, we can see that the youngest student is 17 and the oldest is 42. The most common age group is students in their 20s.

Classroom challenge

1 Kacy recorded the time, in seconds, that it took a group of students to run 400 metres. Her data is shown below.

67	78	79	98	96	103
75	85	94	92	61	80
82	86	90	85	90	89

Copy this stem-and-leaf diagram template and use it to put the information above into an ordered stem-and-leaf diagram.

Step 1: not ordered

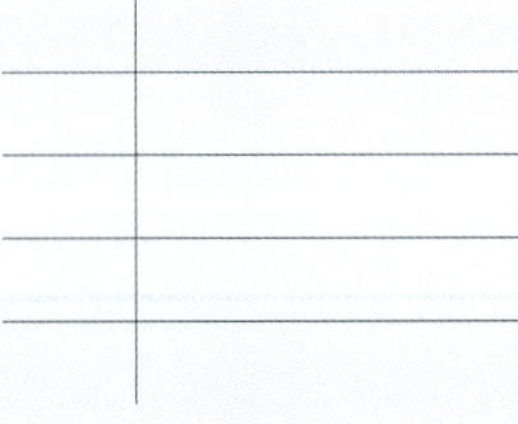

Key:

Step 2: ordered

Key:

2 A mobile police unit recorded the speeds, in miles per hour, of cars travelling along a road.
The data is shown below.

31	52	43	49	36	35	33	29
54	43	44	46	42	39	55	48

Draw an ordered stem-and-leaf diagram to illustrate these speeds.

3 Stephanie noted the time, in minutes, it took her to solve 20 maths puzzles.
Her results are shown below.

5	10	15	12	8	7	20	35	24	15
20	33	15	24	10	8	10	20	16	10

Draw an ordered stem-and-leaf diagram to show these times.

4 Gregory measured the height, in centimetres, of 20 sunflowers he had grown.
The data is shown below.

178	189	147	147	166
167	153	171	164	158
189	166	165	155	152
147	158	148	151	172

Draw an ordered stem-and-leaf diagram to illustrate these heights.

5 The weights, in kilograms, of 15 parcels in an office are shown below.

1·1	1·7	2·0	1·0	1·1	0·5	3·3	2·0
1·5	2·6	3·5	2·1	0·7	1·2	0·6	

Draw an ordered stem-and-leaf diagram to show these weights.

6 The ages of the 15 teachers in the maths department at Eskbank High School are shown below.

35	52	42	27	36
23	31	41	50	34
44	28	45	45	53

Draw an ordered stem-and-leaf diagram to show these ages.

7 Serena was on a geography field trip.
She measured the lengths, in centimetres, of worms she found whilst digging in the earth.
Her results are in the stem-and-leaf diagram below.

Stem	Leaf
1	3 5 7 7
2	0 6 8 8 8 9
3	1 5 5 5 5 6 8 9
4	1 5
5	2

Key: 3|1 = 3·1 cm
N = 21

a What was the length of the shortest worm?
b What was the length of the longest worm?
c How many worms did Serena measure?
d How many worms were 3·5 cm in length?

8 Amir was doing a project for English.
He counted how many words there were in each sentence in an article in a newspaper.
Amir's results are shown in the stem-and-leaf diagram below.

Stem	Leaves
0	8 8 9
1	1 2 3 4 4 8 9
2	0 3 5 5 7 7 8
3	2 2 3 3 6 6 8 8
4	1 2 3 3 5

Key: 1|2 = 12 words
N=30

a How many words were in the shortest sentence?
b How many words were in the longest sentence?
c How many sentences had 27 words in them?
d How many sentences had 'thirty-something' words in them?

9 A fruit grower measured the height, in centimetres, of his tomato plants.
His results are shown in the stem-and-leaf diagram.

Height of Plant (cm)

Stem	Leaves
2	1 3
3	0 2 2 7
4	4 5 6 8 9 9
5	6 7 9
6	3

Key: 3|1 = 31 cm
N=16

a What is the height of the shortest plant?
b What is the height of the tallest plant?
c How many plants are in 'level 4'?
d What fraction of the plants are more than 50 cm tall?

10 Maria recorded the number of emails she received over 21 days.
Her results are shown in the stem-and-leaf diagram below.

Stem	Leaves
0	4 5 5 6 7 7 8 9
1	0 1 2 3 3 4 6 7 8
2	0 1 3 6

Key: 1|2= 12 emails
N = 21

a What was the least number of emails Maria received in a day?
b What was the most number of emails Maria received in a day?
c On how many days did Maria receive 13 emails?
d Write out level 2 in full.

STATISTICS AND PROBABILITY SKILLS

IDEAS OF CHANCE AND UNCERTAINTY

Probability

I can find the probability of a simple event happening and explain why the consequences of the event, as well as its probability, should be considered when making choices. MNU 3 22a

What's coming up?

This Outcome and Experience will give you the opportunity to:
- use the probability scale of 0 to 1 showing probability as a fraction or decimal fraction
- demonstrate understanding of the relationship between the frequency of an event happening and the probability of it happening
- use a given probability to calculate an expected outcome, for example 'the probability of rain in June is 0·25, so on how many days do we expect it to rain?'
- calculate the probability of a simple event happening, for example 'what is the probability of throwing a prime number on a 12-sided dice?'
- identify all of the mutually exclusive outcomes of a single event and calculate the probability of each
- investigate real-life situations that involve making decisions on the likelihood of events occurring and consider the consequences involved.

What you already know

You have already learned how to:
- ✔ use the language of probability accurately to describe the likelihood of simple events occurring, for example: equal chance; fifty-fifty; one in two; two in three; percentage chance; and $\frac{1}{6}$
- ✔ plan and carry out simple experiments involving chance with repeated trials, for example 'what is the probability of throwing a six if you throw a dice fifty times?'
- ✔ use data to predict the outcome of a simple experiment.

Probability scale

Probability describes mathematically how likely it is for an event to happen.

Probability gives us a method of **measuring** the likelihood of an event happening.

You may have heard people say; 'It will *likely* rain tomorrow', or 'United will *probably* win the Cup', or 'I've *no chance* of winning this raffle'.

How does that work?

The **chance** or **likelihood** of an event happening can be shown on a **probability scale**.

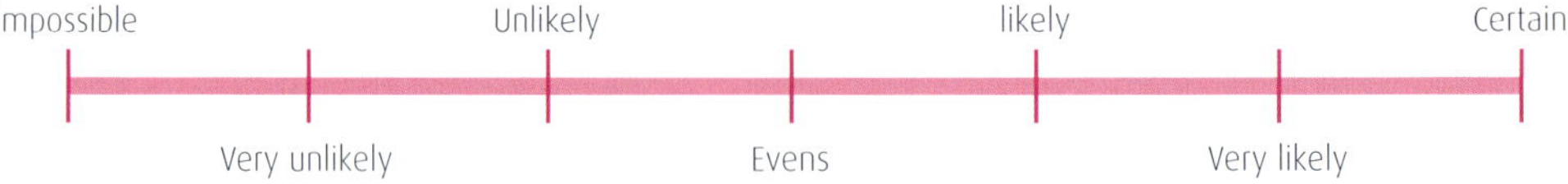

An **even chance** is sometimes seen as '50 : 50'.

You may have heard other ways to describe probability such as 'good chance', little chance', 'no chance', 'highly probable' and so on.

EXAMPLE

Which of the words from the probability line above best describes the chance of the following events happening?

a It will rain tomorrow.
b It will snow tomorrow.
c Tomorrow is Wednesday.
d You will sit down today.
e You will get a 'head' when you toss a coin.

SOLUTION

a This could be *likely* or *unlikely* depending on current weather conditions.
b This will be *unlikely* (unless in the middle of a snowy period).
c If today is Tuesday, then it will be *certain*; any other day and it will be impossible.
d This is *very likely*.
e This is *evens* (that is why a coin is often used to decide who starts a match or chooses 'end').

To measure probability more accurately than the probability line above, we use a numbered probability line. The line goes from 0 (impossible) to 1 (certain), with evens being $\frac{1}{2}$.

Probability is usually written as a fraction, but sometimes as a decimal or as a percentage.

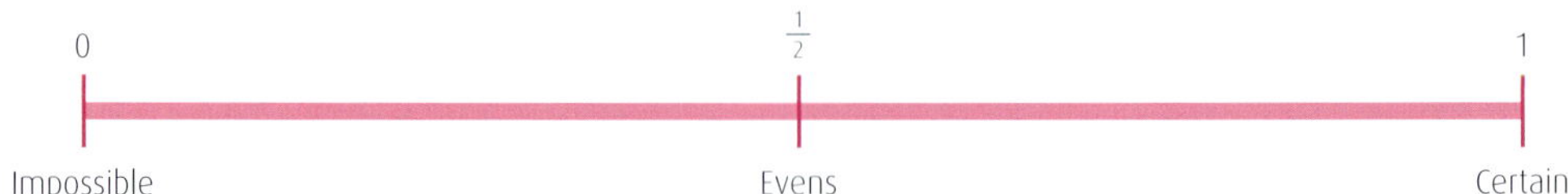

How does that work?

EXAMPLE

Draw a probability line like the one above, and mark on it the probability of the following events happening.

a A person in your class being left-handed.
b Getting a tail when flipping a coin.
c You go to bed before midnight.

SOLUTION

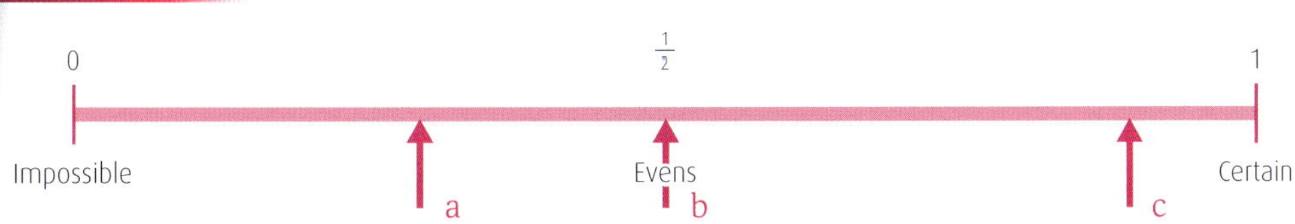

a It is unlikely that there are more left-handed than right-handed people in the class.
b Getting a tail is evens.
c Going to bed before midnight is highly likely.

Classroom challenge

1 Draw a probability line like the one above and mark the following events on it.
   a Your mathematics teacher will give you homework today.
   b It will snow on 14th July.
   c You will miss the school bus tomorrow.
   d You will forget your packed lunch tomorrow.
   e You will be 12 on your next birthday.

2 This probability line shows the probability of five events happening; A, B, C, D and E.

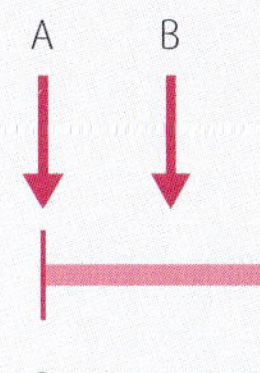

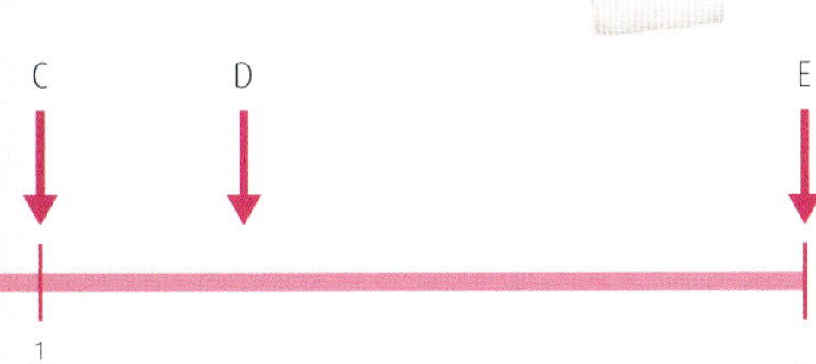

   a Which event is certain to happen?
   b Which event has an even chance of happening?
   c Which event has a chance of happening, but is not very likely?
   d Which event is likely to happen?
   e Which event is impossible?

3 For each of these events, draw a probability line and mark on it the likelihood of the event happening.
  a Scotland winning the next football World Cup.
  b It will rain tomorrow.
  c It will not rain tomorrow.
  d You will get a six when you roll a fair dice (one dice, two dice).
  e You can run a mile in less than 3 minutes.
  f If you roll a dice you will get an even number.
  g You will stand up today.

4 On a probability line mark the probability that you will:
  a live to be 120 years old
  b live to be 100 years old
  c live to be 80 years old
  d live to be 60 years old.

5 a Make a list of events that have an even chance of happening.
  b Make a list of events that have a better than even chance of happening – but are not certain to happen.
  c Make a list of events that have less than $\frac{1}{2}$ chance of happening but are not impossible.

Probability of an event happening

Probability is a **measure** of the likelihood of an event happening.

How does that work?

Consider a fair dice.
What is the probability of rolling a 6?

SOLUTION

We write the probability of rolling a 6 as P(6).
The probability of an event happening is

$$\frac{\text{number of \textbf{successful} outcomes}}{\text{\textbf{total} number of possible outcomes.}}$$

In this case:

There is	1	six in a dice
There are	6	possible outcomes: 1, 2, 3, 4, 5, 6
So	P(6) = $\frac{1}{6}$ (or approximately 0·17)	

On a probability line this would show as:

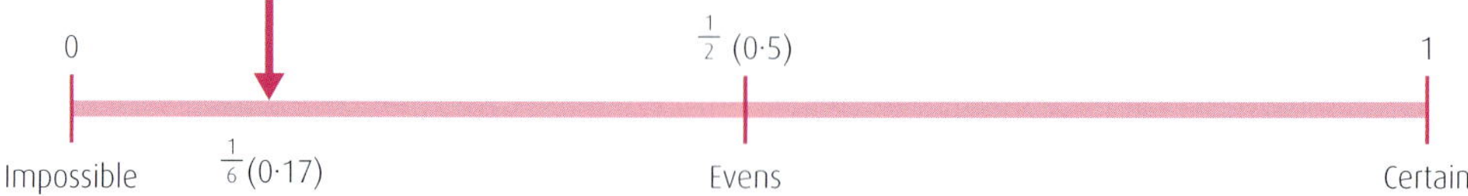

EXAMPLE

What is the probability of picking a card with a heart on it from a pack of cards?

SOLUTION

$$P(\text{Heart}) = \frac{\text{number of cards with a heart}}{\text{total number of cards}}$$
$$= \frac{13}{52} = \frac{1}{4}$$
$$P(\text{Heart}) = \frac{1}{4} \text{ or } 0{\cdot}25$$

EXAMPLE

In a bag there are four blue sweets and six green sweets.

a What is the probability of picking a blue sweet?

b What is the probability of picking a green sweet?

SOLUTION

a $$P(\text{Blue}) = \frac{\text{4 blue sweets}}{\text{10 sweets in total}}$$
$$= \frac{4}{10} = \frac{2}{5} \text{ (or } 0{\cdot}4)$$

b $$P(\text{Green}) = \frac{\text{6 green sweets}}{\text{10 sweets in total}}$$
$$= \frac{6}{10} = \frac{3}{5} \text{ (or } 0{\cdot}6)$$

Classroom challenge

1 Amber is rolling a fair dice.
What is the probability that Amber will roll:
- a a 6
- b an even number
- c a number greater than 4
- d a factor of 6
- e a 7?

2 Catriona is rolling a twelve-sided dice, numbered 1–12.
What is the probability that Catriona will roll:
- a an even number
- b a number greater than 9
- c a prime number
- d a factor of 12
- e a multiple of 4?

3 A spinner has the numbers 0–9 on it. When the arrow is spun, what is the probability the arrow will stop at:
- a an odd number
- b a number less than 3
- c a negative number
- d a prime number
- e a square number?

4 Roland spins a spinner with numbers 1-20 on it. What is:
- a P(7)
- b P(even number)
- c P(odd number)
- d P(number greater than 7)
- e P(factor of 13)?

5 Andre is playing scrabble and has these letters in his rack.
What is:
- a P(W)
- b P(E)
- c P(vowel)
- d P(consonant)
- e P(Z)?

6 A bag contains six red sweets and five yellow sweets. If a sweet is picked at random, what is the probability it will be:

a red
b yellow
c blue?

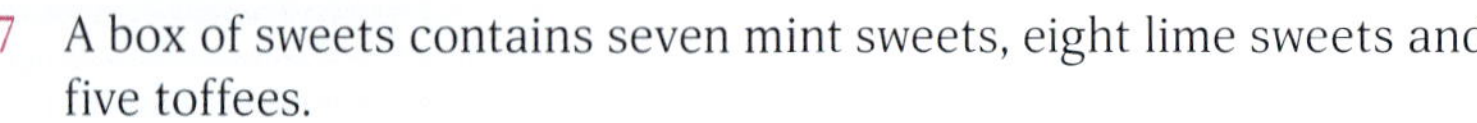

7 A box of sweets contains seven mint sweets, eight lime sweets and five toffees.
What is the probability that a sweet picked at random will be:

a a mint sweet
b a lime sweet
c a toffee
d not a lime sweet
e not a toffee?

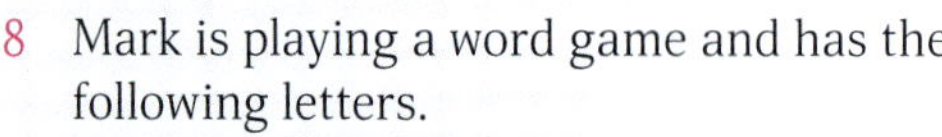

8 Mark is playing a word game and has the following letters.
If Mark picks a letter at random, what is the probability it is:

a an M
b a vowel
c an A
d a consonant
e a Q?

9 An assorted box of doughnuts has three apple doughnuts, two cinnamon doughnuts, four custard doughnuts and three jam doughnuts.
If a doughnut is picked at random from the box, what is the probability it is:

a a cinnamon doughnut
b a jam doughnut
c an apple doughnut
d not a custard doughnut
e a chocolate doughnut?

10 A pack of cards has 13 hearts, 13 diamonds, 13 spades and 13 club cards. For each 'suit', there are cards marked 2–10, Jack, Queen, King and Ace. Hearts and diamonds are red cards, and spades and clubs are black cards.
What is the probability that a card picked at random from the pack is:

a a club
b an ace
c a 'face card' (Jack, Queen, King)
d red
e the 5 of hearts?

Relative and expected frequency of an event happening

How does that work?

The probability of an event happening can be used to estimate the number of times we would expect it to happen over a number of attempts.

EXAMPLE

Kevin flips a coin 50 times. How many 'heads' should he expect to get?
Stacey rolls a dice 120 times. How many 'sixes' should she expect to get?

SOLUTION

The probability of getting a head when a fair coin is flipped is $\frac{1}{2}$.

That is P(head) = $\frac{1}{2}$

In 50 throws, we would expect the number of heads to be $\frac{1}{2} \times 50 = 25$.

If Kevin got a number of heads close to 25, he would expect the coin to be 'fair' or 'unbiased'.

The probability of getting a six when rolling a fair dice is $\frac{1}{6}$.

That is P(6) = $\frac{1}{6}$

Over 120 throws, we would expect the number of sixes to be $\frac{1}{6} \times 120 = 20$.

If Stacey got a number of sixes close to 20, she would expect the dice to be 'fair' or 'unbiased'.

EXAMPLE

Kevin found another coin and wanted to test if it was a fair coin. He flipped it 30 times. The results are shown in the table below.

H	T	H	H	H	T	H	H	T	H
H	H	H	H	T	H	H	H	T	H
H	T	T	H	H	H	H	T	H	H

Is the coin fair?

SOLUTION

Make a table.
The table will show the **relative frequency**. That is, how often each event actually happened.

	Head	Tail
Relative frequency	22	8

Kevin can see that he landed on heads 22 times out of 30.
For a fair coin he would expect the number of heads to be $\frac{1}{2}$ times the number of flips.

Kevin would expect $\frac{1}{2} \times 30 = 15$ heads

Kevin got 22 heads
Kevin might think that the coin was not fair.
However, this was a small 'sample' (see previous chapter), so he may wish to do more flips to confirm his view.

Classroom challenge

1 Sabiha flips a fair coin 100 times.
How many tails would Sabiha expect to get?

2 Jack rolls a fair dice 180 times.
How many sixes would Jack expect to get?

3 Aaron rolls a fair 12-sided dice 240 times.
How often would Aaron expect to see:
a a 12
b an even number
c a square number?

4 Darren spins a nine-sided spinner 108 times.
How often would Darren expect to see:
a a 7
b an odd number
c a prime number?

5 Natasha rolls a dice 30 times. Her results are shown in the table.

3	4	4	4	6	5	4	5	6	4
1	3	2	4	4	5	6	4	4	2
4	5	6	3	2	3	5	6	2	3

Copy and complete this table to show the relative frequency of each number.

	1	2	3	4	5	6
Relative frequency						

Do you think that Natasha has a fair dice or a biased dice?
Explain why you think that.

6 Michael spins the three-sided spinner shown 30 times.
His results are shown in the table below.

A	B	A	C	B	A	A	B	B	C
A	C	A	A	C	C	B	A	A	A
A	A	B	B	C	C	A	A	B	A

Copy and complete this table to show the relative frequency of each letter.

	A	B	C
Relative frequency			

Do you think that Michael's spinner is unbiased or biased?
If Michael spun the spinner 120 times:
a how many As would he expect to get
b how many Bs would he expect to get
c how many Cs would he expect to get?

Making decisions based on chance and uncertainty

How does that work?

EXAMPLE

Alistair and Jenna are in their maths classroom.
Alastair says to Jenna: 'There is a 50:50 chance that the next person through the door is a girl because there are only two choices: a boy or a girl.'
Jenna doesn't think this is necessarily correct.
Can you help them decide if this is correct or not?

SOLUTION

Alastair would be correct if the number of boys and the number of girls in the class was the same.
For example, in a class of 20 students, if there were 10 boys and 10 girls, then:

$P(\text{girl}) = \frac{10}{20} = \frac{1}{2}$ or a '50:50' chance.

But if the class had 12 girls and 8 boys, then:

$P(\text{girl}) = \frac{12}{20} = \frac{3}{5}$

Which is better than 50:50.

So, when making decisions, you need to have an idea of the factors that may affect the probability, or chance, of an event happening – or not happening!

EXAMPLE

Arif says: 'I buy a lottery ticket every week. I am bound to win a big prize sometime.'
Unfortunately, the odds of winning a big prize in the lottery in any given week are about 1 in 14 million!
If Arif lost one week, that does not improve his chances for the next week.

Classroom challenge

1 Write a comment on each of the following statements. Your comment should indicate whether you think the statement is correct or not and give a reason why.

a Craig is flipping a coin. He says: 'The last throw was a head, so the next one will be a tail.'

b Arisa is in her science class. She says: 'The next person through the door will be a boy.'

c Freya says: 'I am unlucky when playing board games as I never manage to roll a 6 on the dice.'

d In a biscuit tin there are chocolate digestives, plain biscuits and ginger biscuits. The probability of picking a ginger biscuit is $\frac{1}{3}$.

e There is a 50% chance that it will rain tomorrow as it will either rain or it will not:

The probability of an event happening = $\frac{\text{number of successful outcomes}}{\text{total number of possible outcomes}}$

2 For each statement below, say whether it is correct or incorrect.

a There are 10 red marbles and 10 blue marbles in a bag. My friend picked out a red marble and then put it back in the bag. I am now more likely to pick out a blue marble.

b Gareth and Eva are rolling a dice. Gareth rolled a 6. Eva says: 'I am not going to get a 6 now!'

c Ian fell off his bike last week. He doesn't fall off his bike very often. He is certain he will not fall off his bike this week.

d Fraser has just flipped a coin three times and got three heads. The probability of getting a head on the next flip is $\frac{1}{2}$.

3 The chance of winning a small prize in a lottery is 0·018. Donald plays the lottery three times a week, every week. He says that he does not expect to win the jackpot, or a big prize, but he does expect to win a small prize regularly.
Do you think Donald is correct?

4 The chance of a pedestrian being killed whilst crossing the road is 1 in 200 000.
Comment on how safe you think it is to cross the road.
Can you list ways that would help to improve the chances of safely crossing the road?

5 The odds of getting a 'hole in one' whilst playing golf is 1 in 5000.
Alan plays 18 holes of golf each week.
Alan thinks he is bound to get a hole in one in the next year.
Do you think Alan is correct? Explain your answer.

6 The odds of being injured by a firework are 1 in 19 566.
Do you think it is safe to go to a fireworks display? Explain your answer.

7 The odds of being injured when mowing a lawn are approximately 1 in 3600.
Geoff mows his lawn once per month, on average. 'Not in a hundred years, will I be injured', says Geoff.
Comment on Geoff's statement.

STRETCH YOURSELF

8 The table shows the likelihood of contracting lung cancer, by the age of 75, for those who start smoking in their teens and then stop at different ages.

Age at which stop smoking	Chance of contracting lung cancer by 75
Never smoked	0·5%
Stop by age of 30	1·8%
Stop by age of 40	3%
Stop by age of 50	6%
Stop by age of 60	10%

Frank thinks that if he stops smoking at 30 he will be 5 times less likely to contract lung cancer than if he kept smoking to 60. Is Frank correct?

Investigate the difference in premiums for life insurance for smokers versus non-smokers.
Does the increase match the table above?

ANSWERS

Number, money, measure: Numeracy skills

Round to 3 decimal places pp9-10

£1 GBP (Great Britain Pound) buys		
Currency	Exchange rate	Written correct to 3 dp
Euro €	1·08011	1·080
US dollar $	1·2937	1·294
Indian rupee ₹	82·6011	82·601
Australian dollar A$	1·6244	1·624
Arab Emirates dirham د.إ	4·75300	4·753
Bulgarian lev лв	2·11725	2·117
Chinese yuan ¥	8·55679	8·557
Egyptian pound ج.م	22·8910	22·891
Hong Kong dollar HK$	10·1189	10·119
Moroccan dirham MAD	12·0645	12·065

Using rounding p11

1

Calculation	Rounded numbers	Estimate answer	Actual answer
18 × 47·3 mm	20 × 50	1000 mm	851·4 mm
12 × £31·15	10 × 30	£300	£373·80
104 × 9·8 g	100 × 10	1000 g	1019·2 g
387 kg ÷ 19·17	400 ÷ 20	20 kg	20·188 kg (to 3 dp)
47·2 + 104·1 + 32·7	50 + 100 + 30	180	184
12121 ÷ 8	1000 ÷ 10	1001	151·51
23·1 × 9·2 + 73·5	20 × 10 +70	270	286·02

2 Approximately 30 × 8 = 240 Actual (232)
3 Approximately 2400 ÷ 8 = 300 Actual (302)
4 Approximately 6 + 10 + 0 + 4 = £20 Actual (£20·25)
5 Approximately 60 + 60 + 50 + 50 + 30 = 250 m Actual (250·85 m)

Recalling my times tables p13

1

Set 1	Set 2	Set 3	Set 4	Set 5	Set 6
3 × 4 = 12	6 × 2 = 12	8 × 4= 32	10 × 3 = 30	12 × 3 = 36	1 × 1 =1
4 × 4 = 16	6 × 8 = 48	8 × 7 = 56	11 × 5 = 55	12 × 6 =72	2 × 2 =4
5 × 3 =1 5	7 × 4 = 28	8 × 12 = 96	5 × 11 = 55	12 × 8 = 96	3 × 3 =9
4 × 7 = 28	7 × 7 = 49	8 × 6 =48	10 × 9 = 90	12 × 5 = 60	4 × 4 =16
3 × 10 =30	7 × 10 = 70	9 × 9 = 81	9 × 10 = 90	12× 10 =120	5 × 5 =25
4 × 11 = 44	6 × 12 = 72	9 × 11 = 99	10×11=110	12 × 7 =84	6 × 6 =36
5 × 12 = 60	7 × 11 = 77	9 × 5 = 45	11×12 = 132	12×12 =144	7 × 7 = 49
4 × 12 = 48	6 × 10 = 60	9 × 7 = 63	10×12 = 120	12 × 2 =24	8 × 8 = 64
3 × 11 = 33	6 × 11= 66	9 × 12 = 108	11×10 = 110	12 × 1 = 12	9 × 9 = 81
5 × 11 = 55	7 × 12 = 84	8 × 11 = 88	11×11 = 121	12 × 0 = 0	10×10 = 100

2

Set 1	Set 2	Set 3
72 ÷ 9 = 8	9 × 4 = 36	144 ÷ 12 = 12
8 × 7 = 56	36 ÷ 9 = 4	10 × 10 = 100
44 ÷ 4 = 11	42 ÷ 7 = 6	88 ÷ 11 = 8
36 ÷ 4 = 9	6 × 7 = 42	48 ÷ 12 = 4
5 × 9 = 45	5 × 7 = 35	55 ÷ 5 = 11
10 × 6 =60	35 ÷ 7 = 5	5 × 11 = 55
48 ÷ 8 = 6	6 × 11 = 66	121 ÷ 11 = 11

Working with numbers in familiar contexts p15

1 £4·75 2 150 km 3 4 kg
4 5 cupboards 5 1260 km 6 1888 passengers
7 11 shelves 8 2 400 000 km 9 £792
10 £756 11 £150 12 $10 400 000
13 125 ml 14 £5760 15 2414 cm
16 1512

Adding and subtracting whole numbers and decimals (up to 3 decimal places) pp17-19

1 a 10·113 b 12·249 c 4·983 d 11·859 e 19·136 f 5·119
g 11·924 h 8·09 i 7·355
2 a 5·385 b 1·385 c 3·935 d 4·281 e 5·754 f 4·987
3 a 8·546 b 7·934 c 6·683 d 4·919 e 7·333 f 9·309
4 £39·64 5 84·065 kg 6 1·79 kg
7 14·951 km 8 36·925 km 9 0·05 m
10 3·488 kg 11 1·565 m 12 5·8 cm
13 0·463 cm 14 0·18 s
15 a 0·896 s b 0·227 s c 1·251 s
d 0·101 mph e 2·145 mph f 3·187 mph
g Valtteri Bottas: 1 h 21 min 41·493 s, Kimi Raikkonen: 1 h 22 min 4 s

Multiplying and dividing whole numbers and decimals (up to 3 decimal places) pp21-22

1 a 17·248 b 42·861 c 65·968 d 66·942
2 a 81·84 b 74·64 c 371·1 d 17·996
3 a 43 b 97·8 c 255 d 232·8 e 4328 f 1570
g 1854 h 25136·4
4 26 t 5 Yes (68·04) 6 32·196 t 7 5·625 kg
8 a 9·478 t b 9·54 t c 6·308 t d 9·925 t e 541·6 t f 5955 t
g 11130 t
9 a 5·448 kg b 2270 kg
10 3·484 knots 11 2283·639 mph

p23

1 a 1·073 b 1·061 c 1·372 d 0·702 e 3·105 f 0·901
g 0·202 h 0·709
2 a 1·811 b 0·071 c 0·151 d 0·062 e 0·032 f 7·022
g 4·01 h 24·24
3 2·2 lb 4 1·354 t 5 9·125 km 6 1·125 m
7 0·454 kg 8 4·896 m 9 0·185 kg 10 £21·99
11 1·652 t 12 0·254 g

Adding and subtracting integers pp25-26

1 a −3, −1, 3, 5 b −4, −1, 0, 2, 4 c 9, 5, 20
d £5, −£5 e 1430 m

2

Complete these number patterns		
Set 1	Set 2	Set 3
6 + 3 = 9	8 + 3 = 11	7 + 3 = 10
6 + 2 = 8	8 + 2 = 10	7 + 2 = 9
6 + 1 = 7	8 + 1 = 9	7 + 1 = 8
6 + 0 = 6	8 + 0 = 8	7 + 0 = 7
6 + (−1) = 5	8 + (−1) = 7	7 + (−1) = 6
6 + (−2) = 4	8 + (−2) = 6	7 + (−2) = 5
6 + (−3) = 3	8 + (−3) = 5	7 + (−3) = 4

3 a 4 b 6 c 4 d −1 e −3 f −7
g −8 h −12

4 a 7 b 12 c 20 d 20 e 0 f −150
g −70 h −500

5 a 5 b 11 c 0 d 2 e 0 f 45
g −38 h −18

6

Set 1	Set 2	Set 3
6 − 3 = 3	8 − 3 = 5	7 − 3 = 4
6 − 2 = 4	8 − 2 = 6	7 − 2 = 5
6 − 1 = 5	8 − 1 = 7	7 − 1 = 6
6 − 0 = 6	8 − 0 = 8	7 − 0 = 7
6 − (−1) = 7	8 − (−1) =9	7 − (−1) = 8
6 − (−2) =8	8 − (−2) = 10	7 − (−2) = 9
6 − (−3) = 9	8 − (−3) = 11	7 − (−3) = 10

7 a 10 b 10 c 14 d 11 e −5 f −3
g 2 h 0

8 a 23 b 34 c 60 d 180 e −50 f −650
g −60 h 0

9 a 9 b 23 c 4 d 8 e 66 f 95
g 2 h 8

10 £0 11 30 12 400 ft 13 125°C

Multiplying and dividing integers pp28-29

1

Set 1	Set 2	Set 3
6 × (−3) = −18	(−4) × 7 = −28	(−4) × (−5) = 20
8 × (−5) = −40	(−5) × 5 = −25	(−3) × (−6) = 18
10 × (−4) = −40	(−8) × 9 = −72	(−8) × (−6) = 48
4 × (−7) = −28	(−12) × 6 = −72	(−12) × (−11) = 132
9 × (−5) = −45	(−11) × 4 = −44	(−9) × (−3) = 27

2 a

×	2	−4	−7	5
2	4	−8	−14	10
6	12	−24	−42	30
−4	−8	16	28	−20
−8	−16	32	56	−40

b

×	3	7	−8	−6
12	36	84	−96	−72
−5	−15	−35	40	30
−6	−18	−42	48	36
10	30	70	−80	−60

3 a 4 × (−5) = −20 b −7 × 6 = −42 c (−5) × −7 = 35
d 12 × −12 = −144 e (−11) × (−10) = 110 f (−8) × −8 = 64

4 a 24 b 48 c −60 d −24 e 48 f −64

5 −8°C

6 a

×	1	2	3	4
1	1	2	3	4
2	2	4	6	8
3	3	6	9	12
4	4	8	12	16

b

×	−1	−2	−3	−4
−1	1	2	3	4
−2	2	4	6	8
−3	3	6	9	12
−4	4	8	12	16

7

Set 1	Set 2	Set 3
6 ÷ (−3) = −2	−4 ÷ 2 = −2	(−4) ÷ (−4) = 1
8 ÷ (−2) = −4	−5 ÷ 1 = −5	(−6) ÷ (−3) = 2
10 ÷ (−5) = −2	−8 ÷ 4 = −2	(−8) ÷ (−2) = 4
4 ÷ (−4) = −1	−12 ÷ 3 = −4	(−11) ÷ (−11) = 1
9 ÷ (−3) = −3	−11 ÷ 11 = −1	−9 ÷ −3 = 3

8 a 5 b −3 c −6 d −2 e −5 f 2
g 25 h 5 i −20 j 12 k −11 l 6

9 a (−40) ÷ 5 = −8 b (−35) ÷ 7 = −5 c (−80) ÷ −20 = 4
d 252 ÷ (−6) = −42 e −30 ÷ (−15) = 2 f 100 ÷ −4 = −25

10 −10°F 11 −57°C 12 61°C

13 a 2 b 2 c −1 d −7 e −14 f −72
g 14

14 a −2 b 2 c −2 d 10 e 8 f −2

Common multiples pp32-34

1 a 4, 8, 12, 16, 20, 24, 28, 32, 36, 40, 44, 48
b 3, 6, 9, 12, 15, 18, 21, 24, 27
c 5, 10, 15, 20, 25, 30, 35, 40
d 10, 20, 30, 40, 50, 60, 70, 80, 90, 100

2 a 32, 36, 40, 44, 48, 52, 56, 60
b 27, 36, 45, 54, 63, 72, 81, 90
c 42, 49, 56, 63, 70, 77, 84, 91, 98
d 54, 60, 66, 72, 78, 84, 90, 96, 102, 108

3 a 2, 4, 6, 8, 10, 12, 14, 16, 18, 20, 22, 24, 26, 28, 30
b even numbers

4

	multiple of	from	to
a	5	15	40
b	6	60	84
c	9	36	90
d	10	70	120
e	20	60	140
f	3	42	54

5 a 12 b 10 c 18 d 20 e 56 f 30
g 21 h 60 i 77

6 a 2, 4, 6, 8, 10, 12, 14, 16, 18, 20 b 3, 6, 9, 12, 15, 18, 21, 24
c 9, 18, 27, 36, 45, 54 d 18

7 a 30 b 30 c 24 d 12 e 24 f 60
g 30 h 12 i 28

8 9:21pm 9 2030 (12 years) 10 9:12am (72 mins)
11 90 seconds 12 72 weeks 13 120 days
14 3 Earth years 15 12 Earth years

Common factors pp35-37

1 1, 2, 4, 8 2 1, 2, 3, 6, 9, 18
3 1, 2, 5, 10 4 1, 5, 25
5 1, 2, 3, 4, 6, 9, 12, 18, 36

6 a 1, 7 b 1, 14, 2, 7
c 1, 24, 2, 12, 3, 8, 4, 6 d 1, 22, 2, 11
e 1, 40, 2, 20, 4, 10, 5, 8 f 1, 29
g 1, 32, 2, 16, 4, 8 h 1, 90, 2, 45, 3, 30, 5, 18, 6, 15
i 1, 54, 2, 27, 3, 18, 6, 9 j 1, 45, 3, 15, 5, 9
k 1, 23 l 1, 60, 2, 30, 3, 20, 4, 15, 5, 12, 6, 10

7 All have an even number of factors

8 a 1 b 1, 2, 4
c 1, 3, 9 d 1, 2, 4, 8, 16
e 1, 5, 25 f 1, 2, 3, 4, 6, 9, 12, 18, 36
g 1, 7, 49 h 1, 2, 4, 8, 16, 32, 64
i 1, 3, 9,27, 81 j 1, 2, 4, 5, 10, 20, 25, 50, 100
k 1, 11, 121 l 1, 2, 3, 4, 6, 8, 12, 18, 24, 36, 48, 72, 144

9 a All have an odd number of factors
b Square numbers

10 a 1, 2 b 1, 3 c 1, 5 d 1, 7 e 1, 11 f 1, 13
g 1, 17 h 1, 19 i 1, 23 j 1, 29 k 1, 31 h 1, 37

11 a Each has 1 pair of factors b Prime numbers

12 a 1, 2, 3, 4, 6, 12 b 1, 2, 4, 8 c 1, 2, 4 d 4

13 a 1, 3, 5, 15 b 1, 2, 4, 5, 10, 20 c 5

14 a 2 b 10 c 12 d 12 e 15 f 4
g 19 h 6

15 a 12 b 3 pencils, 2 pens

16 a 16 b 3 red, 2 white

17 a 6 b 3 bus, 2 tram

18 a 9 b 5 green, 4 blue

19 a 15 b 8 green, 6 blue, 3 white

Working with prime numbers pp38-40

2 a 2×5 b 3×5 c $2 \times 3 \times 3$
d 3×7 e $2 \times 2 \times 2 \times 2$ f $3 \times 3 \times 5$
g $3 \times 3 \times 3$ h $2 \times 3 \times 5$ i 2×43
j $2 \times 2 \times 2 \times 3 \times 3$ k $2 \times 2 \times 2 \times 3 \times 5$ l $2 \times 7 \times 7$

4 Bill 41, 43, 47; Ben 11, 13, 17, 19;
Difference of 26 => 43 and 17

5 $28 = 2 \times 2 \times 7$; $60 = 2 \times 2 \times 3 \times 5$

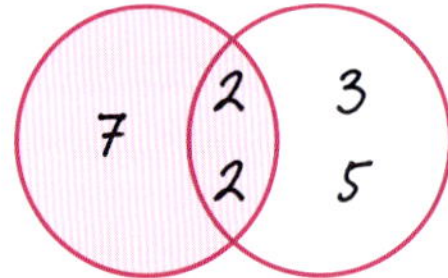

HCF = $2 \times 2 = 4$;
LCM = $7 \times 2 \times 2 \times 3 \times 5 = 420$

6 11, 13, 17, 31, 37, 71, 73, 79, 97

Powers pp42-43

1, 2

Number	Square
1	1
2	4
3	9
4	16
5	25
6	36
7	49

Number	Square
8	64
9	81
10	100
11	121
12	144
13	169
14	196

Number	Square
15	225
16	256
17	289
18	324
19	361
20	400

3 a $3 \times 3 \times 3 \times 3 \times 3 = 243$ b $11 \times 11 = 121$
c $4 \times 4 \times 4 = 64$ d $2 \times 2 \times 2 \times 2 \times 2 \times 2 = 64$
e $5 \times 5 \times 5 = 125$ f 27
g 2401 h 1331
i 46 656 j 1 000 000
k 10 000 l 256
m 1 n 5
o 12 p 512

4 a 2^3 b 4^3 c 5^3 d 10^3 e 100^3

5 $2^{63} = 9{\cdot}3 \times 10^{18}$

Roots pp44

1 Copy and complete this table.

Number	Square Root
1	1
4	2
9	3
16	4
25	5
36	6

Number	Square Root
49	7
64	8
81	9
100	10
121	11
144	12

2 20 m

3 30 cm

4 15 mm

5 Yes as side of frame is 25 cm

Fractions, decimal fractions and percentages

Converting between fractions, decimals and percentages pp47-48

1

Percentage	Decimal fraction	Fraction
50%	0·5	$\frac{1}{2}$
25%	0·25	$\frac{1}{4}$
75%	0·75	$\frac{3}{4}$
$33\frac{1}{3}$%	0·333 (3 dp)	$\frac{1}{3}$
$66\frac{2}{3}$%	0·667 (3 dp)	$\frac{2}{3}$
10%	0·1	$\frac{1}{10}$
20%	0·2	$\frac{1}{5}$
5%	0·05	$\frac{1}{20}$

2

Percentage	Decimal fraction	Fraction
28%	0·28	$\frac{28}{100}$ or $\frac{7}{25}$
30%	0·3	$\frac{3}{10}$
40%	0·4	$\frac{2}{5}$
72%	0·72	$\frac{72}{100}$ or $\frac{18}{25}$
6%	0·06	$\frac{6}{100}$ or $\frac{3}{50}$
85%	0·85	$\frac{85}{100}$ or $\frac{17}{20}$
8%	0·08	$\frac{8}{100}$ or $\frac{2}{25}$
37·5%	0·375	$\frac{3}{8}$
72%	0·72	$\frac{72}{100}$ or $\frac{18}{25}$
80%	0·8	$\frac{80}{100}$ or $\frac{4}{3}$
45%	0·45	$\frac{45}{100}$ or $\frac{9}{20}$
62·5%	0·625	$\frac{5}{8}$
55%	0·55	$\frac{55}{100}$ or $\frac{11}{20}$
1%	0·01	$\frac{1}{100}$
9%	0·09	$\frac{18}{200}$

3 a 0·42 b 0·37 c 0·23 d 0·99 e 0·05 f 0·08
g 0·01 h 0·74

4 a 78% b 26% c 92% d 44% e 3% f 9% g 7% h 12·5%

5 a $\frac{1}{2}$ b $\frac{7}{10}$ c $\frac{11}{50}$ d $\frac{13}{20}$ e $\frac{3}{20}$ f $\frac{1}{25}$ g $\frac{9}{100}$ h 1

6 a 36% b 72% c 48% d 34% e 15% f 95% g 28% h 56% i 80% j 70% k 75% l 10%

7 a $\frac{1}{4}$ b 0·4 c 5%

8 0·0123, 12%, $\frac{1}{5}$, $\frac{3}{8}$, 42%, 0·45

9 9. 45%, $\frac{3}{7}$, 0·35, $\frac{1}{3}$, 30%, 0·04

10 a $\frac{31}{50}$ b 62% c 0·62

11 10%

12 a 75% b 10% c $\frac{15}{100}$ ($\frac{3}{20}$)

13 a 26% b $\frac{26}{100}$ ($\frac{13}{50}$)

14 a 35% b 25% c $\frac{40}{100}$ ($\frac{2}{5}$) d $\frac{15}{20}$ ($\frac{3}{4}$) e $\frac{1}{4}$

Calculate using fractions, decimals and percentages pp50-52

1 a £150 b 6kg c 240m d £12 e 300cm f 5g

2 a £96·20 b 365·5kg c 1025m d £14 e 0·15kg f £81·25

3 a 12 b $\frac{1}{5}$

4 £12 800

5 7

6 a 6·65kg b $\frac{1}{20}$ c 0·25

7 a 6·0g b $\frac{1}{4}$ c 40g

8 a £8 b 11km c 4kg d £8 e 30cm f £600

9 a £18 b 160km c 60 students d £120 e 600 sweets f £4200

10 48 miles

11 60 students

12 33

13 a 216 b 40%

14 a £30 b 37·5%

15 Pets 4U

16 Eatsweet

17 a 400 b 50

18 a 200 b 400 c 150

19 80%

20 7·5%

21 20%

22 Joe

23 80%

24 62·5%

25 37·5%

Multiply fractions p52

1 a $\frac{2}{25}$ b $\frac{3}{8}$ c $\frac{1}{10}$ d $\frac{1}{8}$ e $\frac{6}{35}$ f $\frac{3}{56}$ g $\frac{21}{100}$ h $\frac{8}{27}$ i $\frac{12}{55}$ j $\frac{18}{5}$ k 0 l $\frac{10}{3}$ m $\frac{3}{28}$ n $\frac{1}{2}$ o $\frac{12}{63}$ p $\frac{10}{24}$ q $\frac{1}{12}$ r $\frac{63}{100}$

p53

1 a $\frac{5}{6}$ b $\frac{5}{21}$ c $\frac{14}{33}$ d $\frac{2}{9}$ e $\frac{2}{5}$ f $\frac{1}{4}$ g $\frac{5}{18}$ h $\frac{3}{20}$ i $\frac{3}{10}$

Convert between mixed numbers and fractions pp55-57

1 a $\frac{3}{2}$, $1\frac{1}{2}$ b $\frac{4}{3}$, $1\frac{1}{3}$ c $\frac{11}{5}$, $2\frac{1}{5}$ d $\frac{13}{9}$, $1\frac{4}{9}$ e $\frac{13}{4}$, $3\frac{1}{4}$

2 a $\frac{13}{5}$ b $\frac{11}{8}$ c $\frac{11}{3}$ d $\frac{17}{10}$

3 a $2\frac{1}{5}$ b $1\frac{5}{8}$ c $4\frac{2}{3}$ d $1\frac{1}{10}$

4 a $3\frac{1}{2}$ b $1\frac{2}{3}$ c $2\frac{1}{3}$ d $1\frac{3}{4}$ e $1\frac{3}{5}$ f $2\frac{2}{5}$ g $2\frac{1}{7}$ h $4\frac{2}{7}$ i $6\frac{1}{3}$ j $4\frac{1}{3}$ k $3\frac{1}{10}$ l $10\frac{3}{10}$ m $2\frac{1}{12}$ n $3\frac{4}{11}$ o $4\frac{10}{11}$ p $3\frac{1}{4}$

5 a $\frac{3}{2}$ b $\frac{7}{3}$ c $\frac{9}{5}$ d $\frac{13}{4}$ e $\frac{11}{2}$ f $\frac{11}{3}$ g $\frac{27}{4}$ h $\frac{22}{5}$ i $\frac{33}{4}$ j $\frac{31}{7}$ k $\frac{31}{8}$ l $\frac{38}{5}$ m $\frac{43}{10}$ n $\frac{37}{11}$ o $\frac{25}{12}$ p $\frac{43}{9}$

6 $\frac{18}{5}$, $4\frac{1}{4}$, $\frac{16}{3}$, $6\frac{1}{2}$, $7\frac{1}{3}$

7 $7\frac{1}{4}$, $\frac{13}{2}$, $\frac{23}{5}$, $4\frac{1}{2}$, $2\frac{3}{5}$

8 a = b > c < d > e = f >

9 $4\frac{3}{4}$: 4 × 4 + 3 = 19 => $\frac{19}{4} = \frac{38}{8}$

10 $2\frac{2}{3}$

Add and subtract fractions pp58-59

1 a $\frac{5}{7}$ b $\frac{3}{7}$ c $\frac{3}{4}$ d $\frac{3}{9}$ e $\frac{4}{5}$ f $\frac{4}{5}$ g $\frac{8}{11}$ h $\frac{6}{11}$ i $\frac{4}{5}$ j 1 k 10 l $\frac{37}{50}$

2 a $\frac{7}{10}$ b $\frac{5}{6}$ c $\frac{22}{152}$ d $\frac{8}{11}$ e $\frac{1}{6}$ f $\frac{17}{12}$ g $\frac{9}{20}$ h $\frac{19}{15}$ i $\frac{1}{3}$ j $\frac{1}{8}$ k $\frac{4}{21}$ l $\frac{1}{20}$ m $\frac{17}{21}$ n $\frac{17}{30}$ o $\frac{7}{16}$ p $\frac{1}{16}$

3 a $\frac{3}{8}$ b 3

4 a $\frac{19}{20}$ b $1\frac{1}{20}$km

5 a $\frac{11}{15}$ b $\frac{4}{15}$ c $\frac{1}{3}$

p60

1 a $5\frac{1}{6}$ b $4\frac{1}{4}$ c $2\frac{1}{6}$ d $3\frac{1}{2}$ e $1\frac{13}{20}$ f $1\frac{3}{4}$ g $2\frac{1}{6}$ h $3\frac{17}{20}$ i $4\frac{1}{8}$ j $5\frac{2}{9}$ k $5\frac{1}{2}$ l $2\frac{7}{12}$

2 $\frac{13}{20}$m

3 $\frac{11}{12}$

4 $3\frac{7}{12}$

5 $3\frac{13}{20}$kg

6 $4\frac{11}{12}$

7 $1\frac{3}{4}$kg

8 $1\frac{1}{15}$km

9 $5\frac{13}{20}$kg

10 $2\frac{5}{6}$

11 29 litres

12 $1\frac{9}{20}$

Proportion and ratio

Proportion pp62-64

1 a $\frac{2}{5}$, $\frac{3}{5}$ b $\frac{4}{5}$, $\frac{1}{5}$ c $\frac{3}{10}$, $\frac{7}{10}$ d $\frac{8}{10}$, $\frac{2}{10}$ e $\frac{6}{10}$, $\frac{4}{10}$

2 a r $\frac{3}{5}$, g $\frac{2}{5}$ b r $\frac{2}{4}$, g $\frac{2}{4}$ c r $\frac{4}{6}$, g $\frac{2}{6}$ d r $\frac{1}{5}$, g $\frac{4}{5}$

3

Raspberry concentrate (ml)	Water (ml)	Total volume (ml)	Proportion of raspberry concentrate
10	40	50	$\frac{1}{5}$
20	70	90	$\frac{2}{9}$
45	55	100	$\frac{9}{20}$
60	140	200	$\frac{3}{10}$
15	35	50	$\frac{3}{10}$

4 £3

5 £2

6 £187·50

7 135mins

8 5hrs

9 60

10 20min

11 87·5min

12 3hrs

13 35 days

14 1·4hrs

15 40min

16 1050gal

17 96 days

18 56hrs

19 2·08cm

20 £63·84

Ratio pp66-67

1

Orange squash (ml)	Water (ml)
1	8
10	80
50	400
5	40

2 a 1:2 b 1:3 c 1:5 d 1:5 e 3:1
f 3:1 g 3:1 h 6:1 i 2:3 j 7:3
k 3:4 l 3:2 m 3:7 n 5:3 o 5:8
p 3:1
3 a 2:3 b 2:5 4 1:3 5 3:8
6 a 65:4 b 4:65
7 12 8 70g 9 375cm 10 8
11 a 1·5km b 4km c 6km d 11km
12 32cm

p68

1 a £16, £24 b £10, £25 c 42kg, 18kg d 60cm, 20cm
2 a £10, £20, £30 b £45, £27, £36 c 60kg, 20kg, 40kg
d 151, 271, 331
3 Luke £40, John £50
4 Blue 48ml, Yellow 84ml
5 a 120ml b 15ml
6 A £160, B £120, C £80
7

	a	b	c
Peach	30ml	90ml	150ml
Pineapple	50	150	250
Lemon	10	30	50

8

	a	b	c	d
Yellow	50ml	60ml	60ml	80ml
Blue	100	150	140	80
Red	150	90	100	140

Number, money and measure: Financial skills

Calculate best value in deals and offers pp71–73

1 Gordon's Gourmet Stores
2 Tasko
3 Supermarket
4 1·5litres
5 Ladle
6 300ml
7 School Hoodies (Buy 14 get 7 free)
8 Carry's
9 Ticket Queen
10 (1) Option 2 (2) Option 2 (3) Option 1 (4) Option 1
(5) Option 1 (6) Option 2
11 KC £282, JJ £278, JJ by £4
12 SATtv £226·50, Edge £216, Edge by £10·50
13 a Dawn b Dawn
14 a £41.40 b Yes £33·60 c Lemon cheaper by £1
15 500ml

Effective budgeting and financial vocabulary pp76–77

1 b Deficit £18 c Surplus £42·32 d Balanced £0
2 a Surplus b £219 c 5 months
3 a £162·95 b £118·50 c £44·45
4 £34·78
5 a £1620 b £1180 c £2828
d £1701 e Reduced to £1127
6 a £2910 b £97

Borrowing and credit pp78–80

1 a Offer 2 (for total cost) b Paid off quicker
2 a Loan 2 b Less per month to pay
3 12 monthly payments. Slightly more total cost – but easier to budget monthly payments.
4 £10·50
5

What to compare	Debit card	Credit card
What is it	A debit card is issued by a bank to a customer, who has an account with them. It allows customers to purchase goods or services. The card is linked directly to the bank account	A credit card is issued by a bank (or other financial provide to allow you to buy goods or services 'on credit'. The bank pays on your behalf – and you then repay
When to pay	Taken almost immediately from account	Later, when the statement comes in, a payment is made. Either the full amount, or payments spread over a period of time
Bank Account	A bank account is essential for issuing a debit card	A credit card can be issued without a bank account
Limit	Limit is the amount of funds in the linked account	The maximum amount credit card company allows
Interest	An account may receive interest if in debit	Interest will be charged if full if payment is not made within the agreed time period

6 a i 12% ii £50·40 iii £470·40 iv £39·20
b i £120 ii £210
c i £926·50 ii Yes (£77·21)
d Yes. Total repayment for £1000 is £1090. Total repayment for £1001 is £1071·07
7 Yes as monthly repayments would be £54·60

Converting currencies pp82–84

1 €570 2 $1668 3 ZAR 35427 4 Kr3380
5 A$3045
6 a 3122 b 31136 c $486·50 d €399
7 a 6253 b 1287·60 c 1028·60 d 12483·80
8 Britain 9 £61·40 10 £219·32 11 £100·90
12 a £70·18 b £115·11 c £120·69 d £15·38 e £82·99
f £5·62
13 a €570 b £150 c £131·58
14 a $837 b £602
15 a £11·24 b Approx. £360000 c Approx. £580000000
16 a €285 b €59 c £50
17 a A$3440 b A$440 c £244·44 18 £88·39

Number, money and measure: Measurement skills

Working with hours and minutes pp86

1 a 1·5 b 2·25 c 4·75 d 3·1 e 5·6
f 8·2 g 0·8 h 6·33 i 7·67 j 3·05
k 12·3 l 2·42
2 a 1h 30m b 3h 15m c 6h 45m d 3h 12m e 6h 42m
f 5h 27m g 9h 18m h 12h 47m i 1h 9m j 6h 38m
k 3h 49m l 4h 2m

Calculating time taken for a journey p88

1 a 4h b 4h 24min c 20min d 48min e 6h 42min
f 4h 33min g 7h 18min
2 2h 15min 3 1h 36min 4 2·5s 5 90·6s 6 11·5s
7 7·5min

Calculating average speed on a journey pp90

1 60 km/h 2 12 km/h 3 6 km/h
4 a 63 km/h b 6 m/s c 55 mph d 3400 km/h
e 3·35 m/min (2 dp)
5 48 mph
6 A to B 64 km/h, B to C 64 km/h, C to A 58 km/h
7 a 1 h 20 m b 90 km/h c 72 km/h

Calculating the distance travelled on a journey pp91-92

1 a 60 miles b 180 miles c 330 miles
2 a 6 metres b 30 metres c 60 metres
3 b 225 miles c 56·25 km d 22·5 miles e 3162 miles f 224 km
g 320 km
4 160 cm 5 2700 miles
6 a $\frac{2}{5}$ (0·4) b 26 km
7 a B to M 30 km, M to S 75 km, S to I 279 km b 384 km
c

Brechin			
30 km	Montrose		
105 km	75 km	Stonehaven	
384 km	354 km	279 km	Inverary

8 958 m to the nearest metre

Mixed examples of speed, distance and time pp94-96

1 a 6 h b 120 km/h c 101·25 km d 37·5 miles e 7 h 12 m
f 78 km/h g 220 km
2 5 h 20 m 3 a 09:36 b 65 km/h
4 1820 miles 5 6·55 mph
6 a 720 km/h b 8 h 30 m c 1247 km d i 2 min ii 4 m/s
e 09:00 f i 20 m/s ii 72 km/h
7 No. Average speed is 68·6 mph (Speed limit on motorway is normally 70 mph
8 1·56 km/h 9 14 h 10 3·5 s

Time intervals pp97-98

1 a 2 h 08 min b 5 h 30 min c 6 h 44 min d 11 min e 12 h 35 min
2 a 1 h 50 min b 4 h 07 min c 5 h 33 min d 8 h 20 m e 23 h 58 min
3 20 h 40 min
4 a 8 h 15 min b 8 h 06 min c 9 h 27 min d 11 h 51 min e 16 h 42 min
5 M 5 h 30 min, T 4 h 50 min, W 7 h, Th 5 h 30 min, F 4 h 15 min Total 27 h 05 min
6 a 1 h 40 min b 4 h 35 min
7 AK 103 01:00 EQ 241 02:07 BW 101 23:57 XT 314 02:58
LY 278 02:31

Measurement: Units, areas, volumes

Using appropriate units and converting between units of measure pp102-104

1 a mm b m (cm) c mm d mm² e cm² f cm²
g km h km i ml j ml k litres l cm³
m m² n m³ o cm³ p cm² (m²) q cm r cm³ (l)
s m (or m² for area) t hectares (h)
2 a 5·4 b 23·7 c 7·2 d 24·3 e 200 f 417
g 604 h 1530 i 201·8
3 a 40 b 62 c 35 d 3000 e 1830 f 1000000
4 a 5·19 b 6·08 c 5000 d 8·25 e 4·007 f 3072
5 a 1·765 b 0·432 c 0·025 d 3·482 e 6·062 f 1·5
6 a 4 litres b 0·75 c 35 d 4
7 a 30 000 b 50 000 c 750 000 d 31 410 e 5·4 f 16·72
8 a 400 b 900 c 530 d 4200 e 34·5 f 103·7
9 a 6 b 7·525 c 17·5 d 0·245 e 0·012 f 0·00435
10 a 2000 b 9000 c 4500 d 400 e 370 f 10
11 a 7 b 16·5 c 0·4 d 4000 e 2175 f 1642
12 a 224 b 5400 13 a 107 b 91·4
14 430 000 15 2 500 000
16 a 15 ml b Not quite (13 days)

Area of 2D shapes pp105-107

1 a 21 cm² b 16 m² c 150 mm²
2 a 171 m² b 78 km² c 924 m² d 9 km² e 85 cm²
f 48.825 mm² g 3·915 cm² h 20 cm² i 48·28 mm²
3 a 15600 cm² b 127·5 cm² c 72 cm² d 18·9 m² e 102·9558 m²
f 1500 cm² g 18500 cm² h 52 000 cm²

Volume of a cuboid pp108-111

1 a 120 cm³ b 600 mm³ 2 1200 cm³
3 a 72 000 cm³ b 72 l 4 600 cm³
5 a 192 mm² b cuboid (cube 125 mm³) 6 48 m³
7 a 60 000 cm³ b 60 l 8 480 cm³
9 1053 m³ (just over 1 l which is 1000 cm³)
10 540 m³ = 540 000 l 11 a 700 cm³ b 6 blocks
12 a 30 000 cm³ b 250 cm³ c 6 × 5 × 4 = 120
13 a 400 000 cm³ b 10 500 mm³ c 3·315 m³ d 30 000 cm³
e 8·25 m³ f 864 l g 4·8 m³
14 a 75·6 l b 71·4 l c 42 000 cm³ d 4·2 l

Area of compound shapes pp114-116

1 a 32 cm² b 33 cm²
2 a 34 m² b 180 cm² c 144 cm² d 84 mm²
3 a 4000 cm² b 392 m² c 1125 cm² d 3060 mm² e 160 m²
f 4600 mm²
4 a 71 cm² b 198 m² c 192 cm² d 104 m²
5 a 398 mm² b 98 m² c 36 cm² d 78 cm²
6 a 112·8 m² b Yes; needs 12 tins which costs £300

Volume of compound objects pp118-119

1 400 cm³ 4 690 cm³ 7 168 cm³ 10 1·232 m³
2 17 400 cm³ 5 1450 cm³ 8 1·89 m³ 11 1125 cm³
3 7·2 m³ 6 9 m³ 9 3·744 m³

Number, money and measure: Algebraic skills

Patterns and sequences pp125-126

1

1st Term	Rule	Sequence
1	Add 3	1, 4, 7, 10, 13, 16, … …
3	Double	3, 6, 12, 24, 48, 96, … …
4	Add 6	4, 10, 16, 22, 28, 34, … …
2	Multiply by 5	2, 10, 50, 250, 1250, … …
80	Subtract 5	80, 75, 70, 65, 60, … …
10	Multiply by 10	10, 100, 1000, 10 000, 100 000, … …
6	Add 6	6, 12, 18, 24, 30, 36, … …

2 a 26, 30, +4 b 47, 55, +8 c 75, 72, −3 d 972, 2916, ×3
e 33, 22, −11 f 3125, 15625, ×5 g 45, 36, −9

3 a 25, 30, +5 b 20, 24, +4 c 45, 54, +9 d 50, 60, +10 e 15, 18, +3
f 12, 14, +2 g 35, 42, +7
4 a 35 b 50 c 5000
5 a 28 b 40 c 400 d $4n$
6 a $9n$ b $10n$ c $3n$ d $2n$ e $7n$
7 b 5, 9, 13, 17, 21 c 8, 13, 18, 23, 28 d 5, 11, 17, 23, 29
e 7, 14, 21, 28, 35 f 5, 13, 21, 29, 37 g 8, 18, 28, 38, 48
h 12, 24, 36, 48, 60 i 11, 22, 33, 44, 55 j 1, 4, 7, 10, 13

Sequences and algebraic notation pp127–128

1 a $C = 60 + 30h$ b 90, 120, 150, 180, 210, 240, 270
2 a $C = 2 + 0{\cdot}5w$ b i £4·50 ii £7 iii £9·50
3 a $C = 4 + 1{\cdot}2m$ b £13·60 4 $C = 32n$ 5 $C = 1{\cdot}33l$
6 a £87 b $C = 12 + 25l$ c i £137 ii £262 iii £387
7 a $C = 100 + 45d$ b 145, 190, 235, 280, 325, 370, 415
c £640
8 a 90, 155, 220, 285, 350 b $C = 25 + 65d$ c i £480 ii £935
9 a $C = 50 + 85h$ b £475 c 4 hours

Simplifying expressions pp130–131

1 a $2t$ b $4t$ c $3d$ d $5k$ e $6a$ f $3x$
g $3g$ h $2x$ i 0 j $3t$ k $2n$ l w
2

a	$t + t + t + t + t$	$5 \times t$	$5t$
b	$m + m + m$	$3 \times m$	$3m$
c	$w + w + w + w + w + w + w + w$	$8 \times w$	$8w$
d	$x + x + x + x + x + x + x$	$7 \times x$	$7x$
e	$k + k + k + k + k + k$	$6 \times k$	$6k$
f	$p + p + p + p + p + p + p + p + p + p$	$10 \times p$	$10p$
g	$y + y + y + y + y + y + y + y + y$	$9 \times y$	$9y$

3 a $4m$ b $6k$ c $9t$ d $2r$ e $3k$ f $6z$
g xy h ab i $\frac{n}{3}$ j $\frac{t}{6}$ k $\frac{7}{w}$ l $\frac{6}{m}$
m qt n w^2 o s^2 p a^2
4 a $3ab$ b $4mn$ c $6pq$ d $15cd$ e $6m^2$ f $12k^2$
g $24mn$ h $24abc$

Collecting like terms p132

1 a $6x$ b $8k$ c $4t$ d $2m$ e $6k$ f $4m$
2 a $5m$ b $6x$ c $7k$ d $4p$ e $6a$ f y
3 a $7m + 2$ b $5x + 7$ c $9k + 4$ d $2b + 2$ e $8z - 4$ f $5t - 4$
4 a $5m + 7k$ b $5x + 4y$ c $4n + m$ d $t + 3s$ e 0 f $11w + 4v$
5 a $11ab$ b $9w^2$ c $12xy$ d $2mn$ e $-9k$ f $9y$
6 a $10m + 5g$ b $5b + c$ c $7f + g$ d $5d^2 + 10de$
7 a $5a + 6b + 9c$ b $5m + 7n + 9k$ c $x + 5y + 7z$ d $10d + 2e + 5f$
e $6u - w + 3v$ f $2t - 3r - 2u$ g $5q$ h $-14h$
i $5a^2 + 5a$ j $2w^2 + 6vw$

Substitution p134

1 a 7, 11, 3 b 5, 23, 1 c 19, −2, 33 d 27, 21, 0 e 3, 0, 21
f 50, 0, 125 g 18, 40, −2 h 4, 0, −14
2 a 8 b 21 c 29 d 13 e 3 f 16
g 49 h 65 i 33
3 a 20 b −12 c 17 d 8 e 18 f 36
g 25 h 50 i 9 j 34 k 59 l 0
4 a 8 b 15 c −10 d 11 e 43 f −30
g −6 h −1 i −11 j 5 k −36 l 50

5 a 16 b 25 c 41 d 29 e 21 f 100
g 80 h 9 i 98
6 a 36 b 4 c 40 d 32 e 144 f 144
7 a 4 b 9 c 8 d 2 e 4 f 12

Evaluating a formula pp136–138

1 120 cm² 2 14·4 m² 3 30·48 cm 4 99·06 cm
5 a 86 °F b 212 °F c 32 °F d −40 °F
6 a 40 °C b 100 °C c 0 °C d −40 °C
7 a £90 b £150 c £110 d £140
8 a £80 b £155 c £67·50 d £92·50
9 a 80 b 180 c 32·5 d 6·25
10 a 300 cm³ b 1·5 m³ c 396 mm³ d 7500 mm³
11 1800 000 000 000
12 a 400 ft b 6400 ft c 1024 ft d 676 ft
13 a 150 cm² b 384 m² c 73·5 cm² d 132·54 m²
14 132 m 15 57·8 m

Creating a formula: simple linear patterns pp139–141

1 a

Squares (s)	1	2	3	4	5
Lollipop sticks (l)	4	8	12	16	20

b 4 c $l = 4s$ d 32
2 a

Hexagons (h)	1	2	3	4	5
Lollipop sticks (l)	6	12	18	24	30

b 6 c $l = 6h$ d 42 e 90
3 a

Octopuses (p)	1	2	3	4	5
Tentacles (t)	8	16	24	32	40

b 8 c $t = 8p$ d 160
4 a

Textbook (t)	1	2	3	4	5
Cost (c)	20	40	60	80	100

b 20 c $C = 20t$ d £600
5 a $f = 7v$ b $m = 110c$ c $t = 15m$ d $p = 9t$ e $m = 60h$

More linear patterns pp143–146

1 a

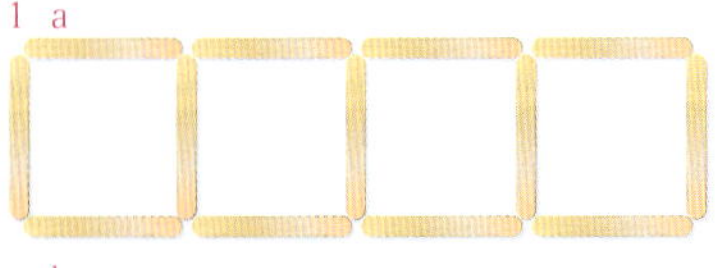

b

No of squares (s)	1	2	3	4	5	6
No of lollipop sticks (l)	4	7	10	13	16	19

c 3 d $l = 3s + 1$ e 46 f 20 g 20
2 a

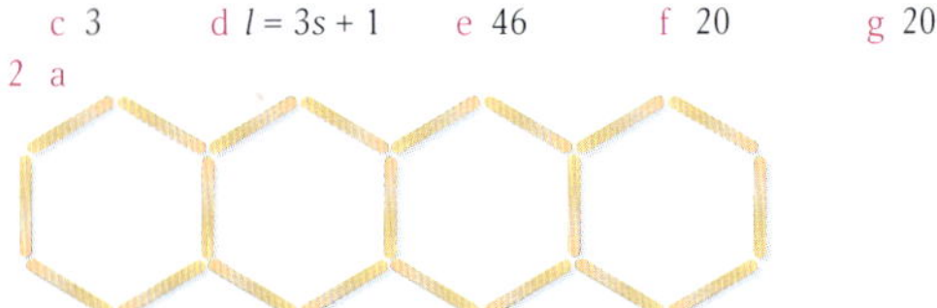

b

No of hexagons (h)	1	2	3	4	5	6
No of sticks (s)	6	11	16	21	26	31

c 5 d $s = 5h + 1$ e 51 f 126 g 18

3 a

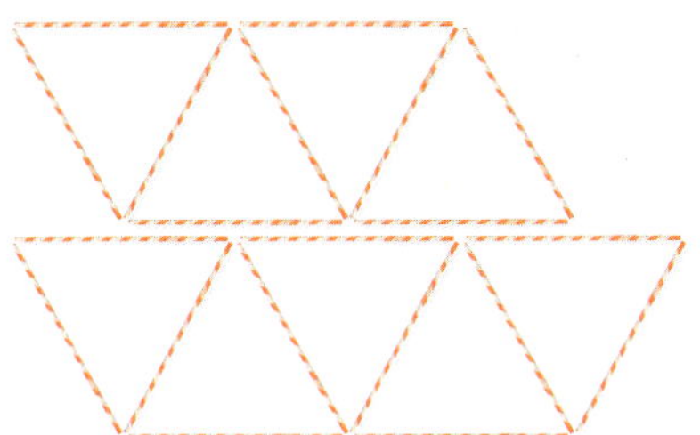

b

No of triangles (t)	1	2	3	4	5	6
No of straws (s)	3	5	7	9	11	13

c 2 d $s = 2t + 1$ e 17 f 25 g 50

4 a

No of days hired (d)	1	2	3	4	5	6
Cost in £ (c)	9	19	29	39	49	59

b £10 c $C = 10d - 1$ d i £99 ii £139 iii £209 e 8 days

5 a 0·5 b $w = 0{\cdot}5g - 0{\cdot}1$ c 8·9 l

d Yes. 2·4 litres are needed

6 a

b

Pattern (p)	1	2	3	4	5	6
Number of bricks (b)	5	9	13	17	21	25

c $b = 4p + 1$ d 37 e 12th

7 a $f = 2d + 1$ b $t = 20p + 20$ c $b = 5p + 7$ d $c = 15d - 3$

e $c = 10b - 2$

Creating a formula from a diagram or problem pp148–149

1 a $P = 2a + 2b$ b $P = l + m + n$ c $P = p + q + 2r + 2s$

2 a $A = ab$, 60 cm² b $A = a^2$, 9 cm² c $A = a^2 + ab$, 20 cm²

d $A = ab + bc$, 18 cm²

3 a $n + 1, n + 2$

b Total $= n + n + 1 + n + 2 = 3n + 3$ or $3(n + 3)$ which is divisible by 3

4 $C = 2p - 1$, 15

5 $n + n + 5 = 27$, $2n + 5 = 27$, $n = 11$ years old

6 a

b $C = 2m + 6$

Equations p151

	a	b	c	d	e	f
1	a 3	b 6	c 5	d 11	e 12	f 18
	g 2	h 0	i −1	j −2	k 1	l −3
2	a 4	b 3	c 7	d 3	e 3	f 20
	g 4	h 8	i 11	j 6	k −3	l −8
3	a 5	b 4	c 4	d 3	e 4	f 2
	g 5	h 5	i 5	j 3	k 4	l 0
	m 5	n 8	o 7	p 2·5	q 3·5	r 3·5
4	a 3	b 4	c 7	d 5	e 1	f 3
5	a −2	b −2	c −1			
6	a 12	b 12	c 14	d 15	e 18	f 0

Puzzle: A + D + E = 5B so $k = 5$

Shape, position and movement: Shape

Drawing triangles pp158–159

1–4 Sides should be drawn to within ±2 mm and angles within ±2°

5 The sides of length 4 cm and 5 cm cannot form a triangle with a side of 10 cm as $4 + 5 = 9 < 10$ so there would be a gap.

Constructing other 2D shapes: regular polygons p160

1 a 5 b 360

Constructing other 2D shapes p161

1 b 130° c PQ 3·8 cm ±2 mm, QR 5·8 cm ±2 mm

Shape, position and movement: Angles, symmetry and transformations

Naming angles pp164–165

1 a ∠BAC, ∠CAB b ∠QPR, ∠RPQ c ∠RST, ∠TSR

d ∠LMN, ∠NML e ∠TSU, ∠UST f ∠XYZ, ∠ZYX

2 a ∠ADB, ∠CDB b ∠MLJ, ∠JLK c ∠ECD, ∠DCF

d ∠SPR, ∠QPR e ∠BAD, ∠CAD f ∠BEC, ∠DEC

3 a ∠ABC, ∠CBD, ∠ABD

Parallel lines cut by a third line pp166–168

1 a $a = 50°$ b $b = 35°$ c $c = 20°$

2 a 120° b $b = 90°$ c $c = 60°$

3 a $a = 120°$ b corresponding

c $b = 65°$ d $c = 135°$, $d = 45°$

4 a $a = 144°$ b vertically opposite

c $b = 56°$ d $c = 137°$, $d = 43°$

5 a $a = 43°$ b alternate c $b = 62°$ d $c = 123°$, $d = 57°$

6 a $a = 37°$, corresponding, $b = 37°$, vertically opposite

b $c = 121°$, alternate, $d = 59°$, supplementary to c

c $e = 53°$ corresponding, $f = 53°$, vertically opposite to e, $g = 127°$, supplementary

d $l = 92°$, corresponding, $j = 88°$, supplementary to 92° $k = 92°$, vertically opposite to l, $h = 92°$, alternate to i

e $l = 101°$, corresponding, $m = 79°$, supplementary $n = 101°$ vertically opposite to l, $p = 79°$, vertically opposite to m

f $r = 26°$, vertically opposite, $s = 154°$, supplementary, $q = 154°$, vertically opposite to s, $u = 154°$, corresponding, $v = 26°$, corresponding, $w = 154°$, vertically opposite $t = 26°$, alternate g $x = 48°$, supplementary, $y = 48°$, corresponding, $z = 132°$, corresponding

h $a = 50°$, vertically opposite, $b = 70°$, supplementary to 50 + 60, $c = 70°$, alternate to b, $d = 50°$, $e = 130°$, alternate to 60 + 70 (b) $f = 110°$, alternate to $a + 60$, $g = 110°$, vertically opposite to f, $h = 70°$, vertically opposite to c

7 $b = 37°$, supplementary, $c = 37°$, alternate, $a = 180 - 37 - 37 = 106°$, isosceles triangle

Angles in a triangle pp170–171

1 a $r = 64°$ b $e = 63°$ c $z = 32°$ d $h = 30°$ e $a = 60°$ f $x = 119°$
g $k = 30°$ h $v = 66°$ i $t = 65°$, isosceles j $k = 60°$, equilateral

2 a 72·5°, 72·5° b 39°, 39° c 42°, 96°

3 a $a = 37°$, $b = 37°$ b $c = 23°$, $d = 134°$, $e = 46°$

Angles in a quadrilateral p172

1 a 89° b 52° c 84° d 65° e 106° f 96°
g 106° h 67° i 99° j 111° k 96°

Bearings and scale: Bearings

How to draw a bearing pp176–177

1 a 150° b 095° c 127° d 060° e 238° f 315°

2 a 105° b 285°

3 a 078° b 258° c 125° d 040° e 260°

4 Practical

5 a 080° b 315° c 030° d 330° e 235° f 010°
g 082° h 115° i 175° j 100° k 125° l 250°
m 175° n 015°

Using scale and bearings pp179–181

1 b 220m c 235° 2 b 132° c 570m

3 b 225km 4 70m, 115°

5 a 290° b 31·5km c/d See diagram below.

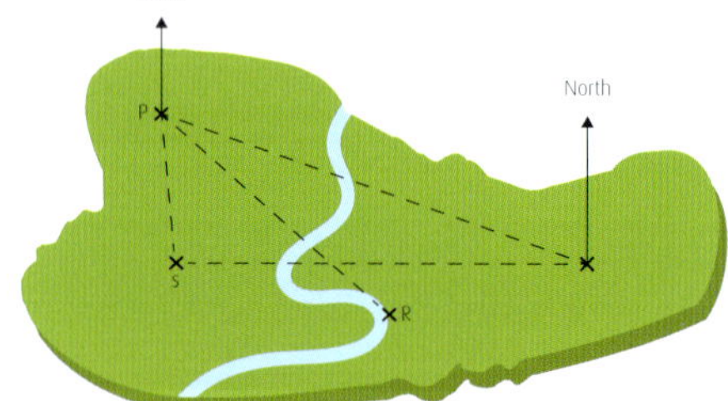

6 a 24km b See diagram below

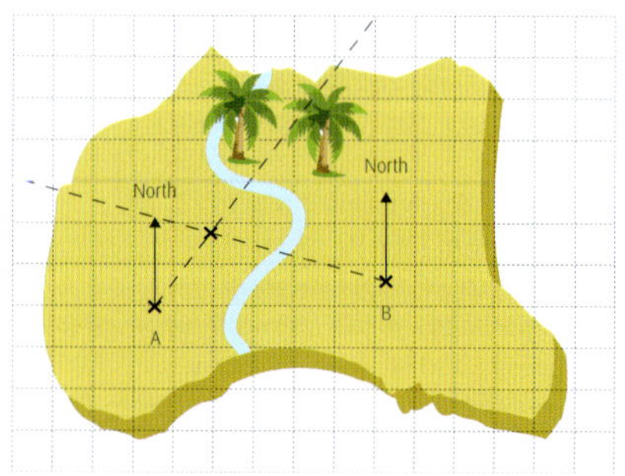

7 a See diagram below b 28km c 280°

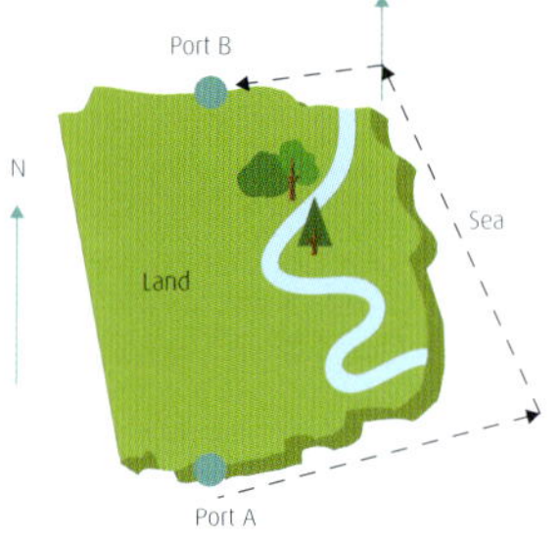

8 555m

9 a Students' scale drawings b 415km c 265°

pp 182–185

1 1·5km 2 1·85km 3 8km 4 10·8km

5 a 340°, 3·6km b 260°, 10km c 300°, 4km d 055°, 7km
e 085°, 6km

6 6cm 7 15cm 8 4·33cm

9 a 3·5cm b 1·5cm 10 a 7·5cm b 50000000km

11 M 0·58cm, V 1·08cm, E 1·5cm, M 2·28cm, J 7·8cm, S 14·3cm, U 28·7cm, N 45cm, P 60cm

12 a 115km b 400mm c 380mm d 150km e 360mm f 275mm

Enlargement/reduction pp187–189

1

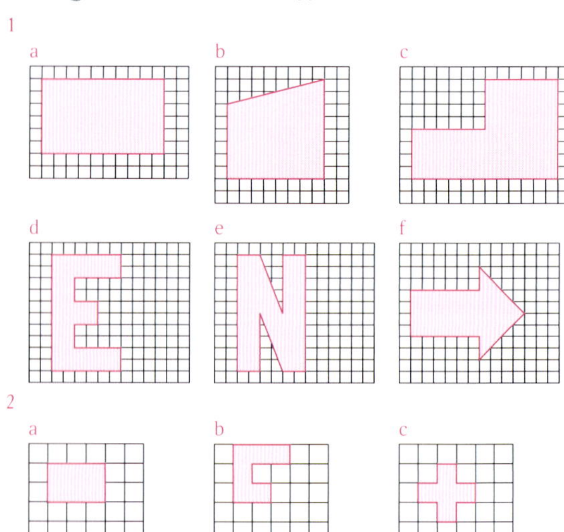

3

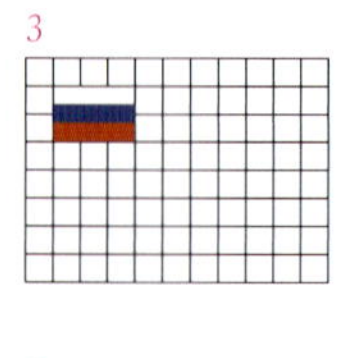

4

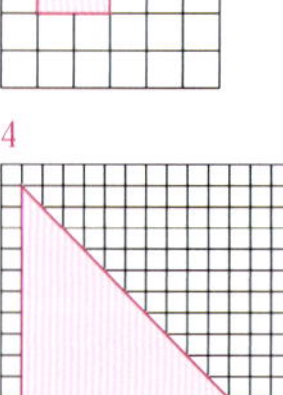

5

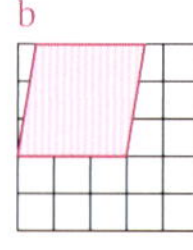

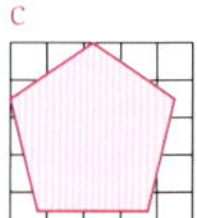

Enlarge or reduce shapes pp191-193

1 a 7:420, 1:60 b 9 cm c 1·5 m
2 a 24 m b 3·75 m
3 a 1:500 b 80 m c 1200 m^2
4 a l = 8 cm, b = 3 cm b l = 16 cm, b = 6 cm
c l = 4 cm, b = 1·5 cm d l = 80 cm, b = 30 cm
5 a 1:50 b 180 cm c 60 cm
6 a l = 4·8 m, w = 3·6 m b yes
7 Practical 8 Practical

Using a protractor and ruler to make a scale drawing p194

1 16·8 m 3 11·7 m 5 457 m
2 36·4 m 4 107·2 m 6 1164·7 m

Coordinates pp196-198

Plot in all four quadrants
1 A(1, 6) B(1, 3) C(6, 3) D(6, 6)
2 b Square
3 A(2, 5) B(4, 5) C(4, 3) D(2, 1) E(1, 2) F(2, 3) G(1, 4)
4 Shape a Square Shape b Rhombus
Shape c Parallelogram Shape d Triangle
Shape e Rectangle Shape f Pentagon
Shape g Trapezium Shape h Octagon
5

Shape	Coordinates
Triangle	(2, 1)
Cross	(10, 4)
Heart	(10, 3)
Square	(9, 8)
Lightning	(1, 9)

Shape	Coordinates
Circle	(7, 8)
Diamond	(0, 8)
Star	(7, 7)
Music Note	(0, 1)
Moon	(4, 9)

6 d (2, 7) and (8, 1)
7 b Parallelogram, 30 cm^2 d Isosceles 24 cm^2
f Trapezium, 20 cm^2 h Rhombus, 12 cm^2
8 a I am finished
b (0, 6), (1, 9), (0, 8), (8, 3), (2, 0) c (0, 8), (1, 9), (9, 2), (2, 0)
d (0, 8), (1, 9), (3, 4), (5, 7), (9, 2), (2, 0)
e (0, 8), (8, 3), (3, 4), (1, 9) f (8, 6), (1, 9), (2, 0), (8, 3), (2, 0)
g (5, 4), (8, 3), (3, 4), (5, 7), (9, 2), (6, 3)

Bilateral symmetry pp200-202

1 a Yes b Yes c No d No
2

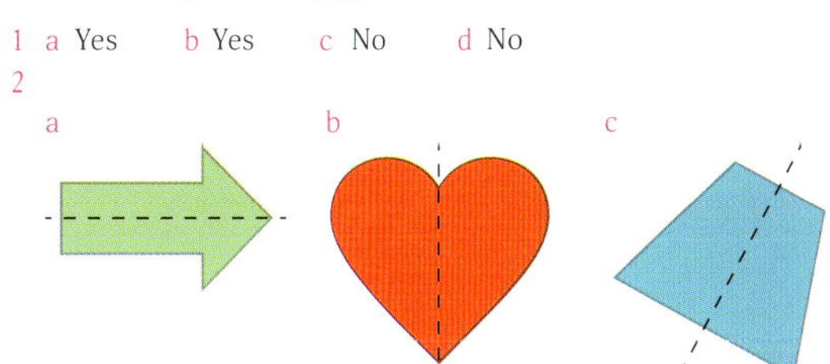

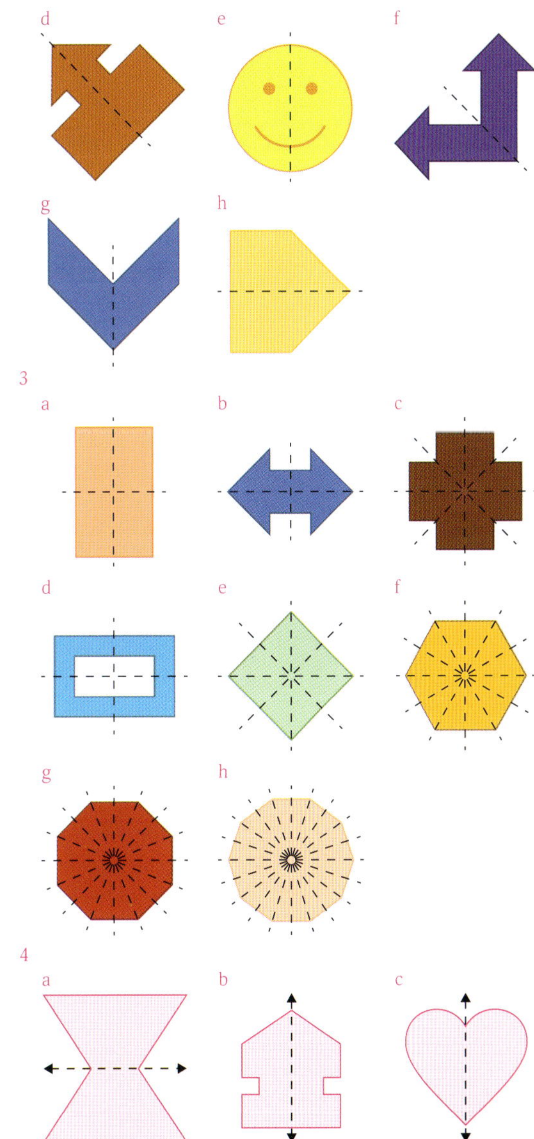

5

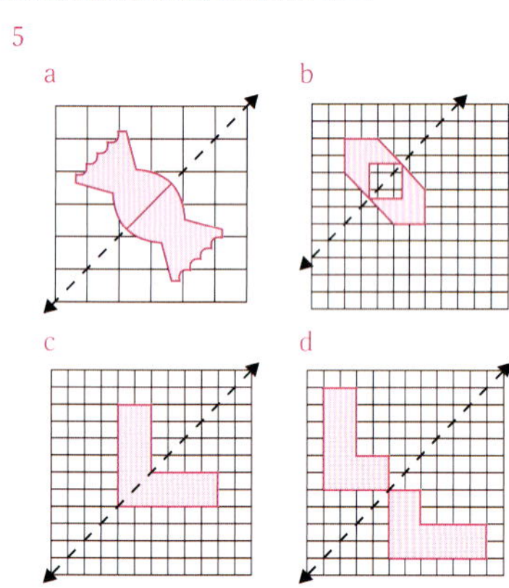

Information handling: Data handling skills

Collect, interpret and describe trends in data
Working with charts and tables pp208-211

1 a 165 b Choi c 57 d i increasing ii decreasing

2 a 169 b 18 c Wednesday; a lot fewer sales d Cream

3 a 15m b 5m c 11m d 96m e Yes (36 < 40)

4 a 8·41m b 2nd c 0·19m d 0·19m

5 a £357 b £426 c £417 d £532
e 2nights, Jan–Aug and 3 nights Jan–Apr

6 a £89 b 9h 15min c £298
d Flight times are a lot longer

7 a £1·95 b 10% c £3·25 d £7·80

8 a approx. 133cm b 16 years c 16–17
d slowly increasing (rising)

9 a 4 b 8 c January d May
e Decreasing (downward)

10 a Obese b Normal c Overweight d about 17–18lbs

Working with graphs pp213-216

1 b Comparative bar chart (p215)

2 a

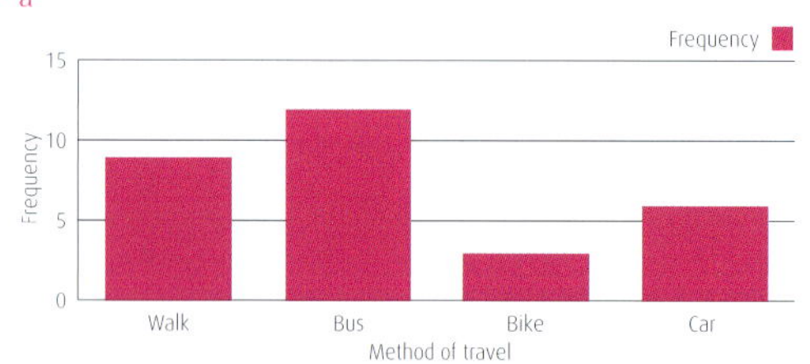

b Sectors are: walk 108°, bus 144°, bike 36°, car 72°

3 a See diagram below

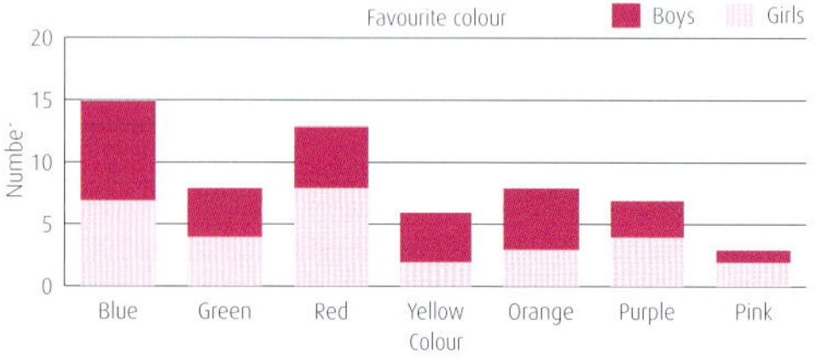

b See diagram below

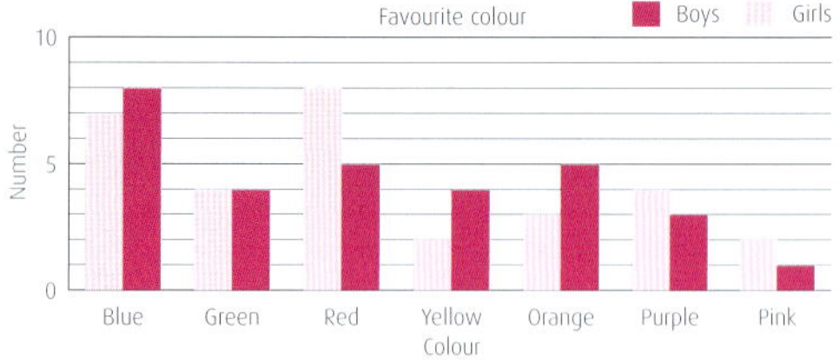

c Blue d Red e Green

4 a A b A c 53·8% d 37·5%

5 a Week 2 b Approx. 46 c 27% d Type B
e Type C – fewest sales

6 a 47 b $\frac{17}{47}$ c 7 d 9%

7 a 4 b 5 c 1 d 18 e 33·3%

8 a Bus b Yes, angle at centre of car drivers sector is about 160° and angle at centre of car passengers is about 80°.
c about $\frac{1}{3}$

Bias and sample size pp217-220

1 Graph 2, because the scale does not start at 0

2 Vertical scale does not go up in same increments (first increment = 100, then other increments = 50)

3 Adjust vertical scale (e.g. start at 0 and then have increments of 10)

4 a Yes b Adjust 0–10 gap so same as others

5 Graph 2, does not start at 0

6 a Graph 2 b Yes, graph does not start at 0 and scale 'skews' increase to look greater

7 Vertical scale inconsistent (0–6, 6–10, 10–16, 16–20)

8 5 equal sectors so expect about 10 of each in 50 throws. Most ok but possible high 'reading' for 4

9 a Biased (attending a particular sport b Unbiased (random)
c Biased (attending a particular group)
d Unbiased (ask all population)

10 a Use only one year group
b Select same sample number from each year group

Pie charts p221-223

1

Blue	Red	Yellow	Other
63°	180°	72°	45°

2

Mathematics	English	French	Art
144°	108°	36°	72°

3

Dog	Cat	Gerbil	Snake	Other
144°	84°	72°	12°	48°

4

White	Red	Blue	Black	Silver
90°	66°	72°	102°	30°

5

Blue	Green	Hazel	Grey	Brown	Other
120°	66°	54°	45°	54°	21°

6

Cow	Goat	Pig	Sheep	Chicken
180°	24°	50°	68°	38°

7

Ford	WV	Suzuki	Fiat	Renault
72°	108°	108°	12°	60°

8 a Food b $\frac{1}{4}$ c £120 9 a Walk b 6 c 8

10 a 30 b 10

c No. Do not know how many medals, in total, Ghana won

Line graphs p223-226

1 a 12 b 2016 c 2013 and 2014 d 2

e Increasing

2 a 60°F b 95°F c May and Sept d Decreasing/falling

e 10°F

3

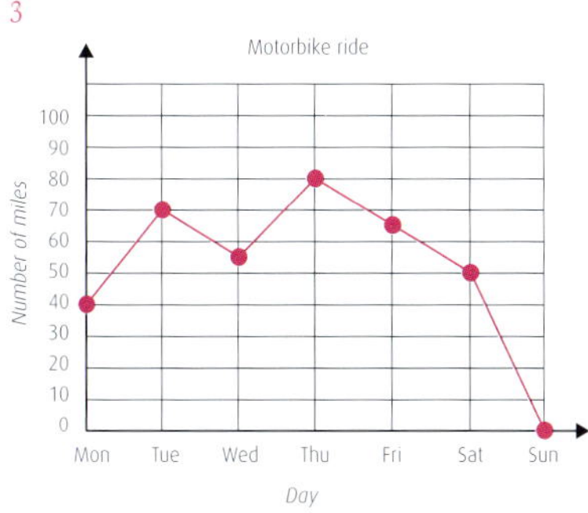

4

5

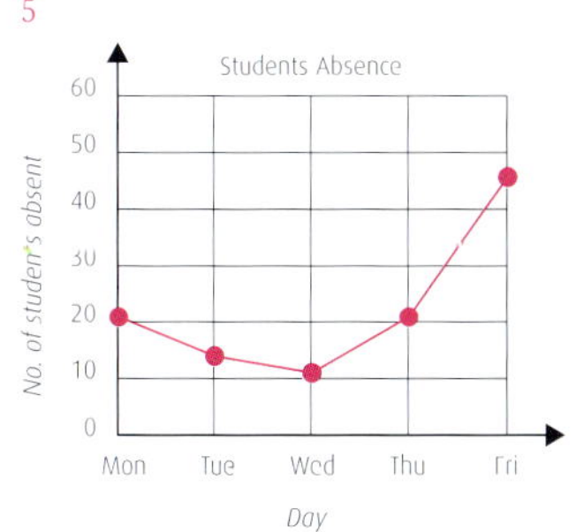

6 a 2016 b 225 c 2016 and 2018

d 2013 and 2017 e 2015–16

7

Stem and leaf diagrams pp228-230

1

6 | 1 7
7 | 5 8 9
8 | 0 2 5 5 6 9
9 | 0 0 2 4 6 8
10 | 3

KEY
6|1 = 61 seconds
N = 18

2

2 | 9
3 | 1 3 5 6 9
4 | 2 3 3 4 6 8 9
5 | 2 4 5

KEY
3|1 = 31 mph
N = 16

3

0 | 5 7 8 8
1 | 0 0 0 0 2 5 5 5 6
2 | 0 0 0 4 4
3 | 3 5

KEY
2|0 = 20 mins
N = 20

4

14 | 7 7 7 8
15 | 1 2 3 5 8 8
16 | 4 5 6 6 7
17 | 1 2 8
18 | 9 9

KEY
16|8 = 168 cm
N = 20

5

0 | 5 6 7
1 | 0 1 1 2 5 7
2 | 0 0 1 6
3 | 3 5

KEY
1|3 = 1·3 Kg
N = 15

6

2 | 3 7 8
3 | 1 4 5 6
4 | 1 2 4 5 5
5 | 0 2 3

KEY
3|5 = 35 years
N = 15

7 a 1·3 cm b 5·2 cm c 21 d 4

8 a 8 b 45 c 2 d 8

9 a 21 cm b 63 cm c 6 d ¼

10 a 4 b 26 c 2 d 20, 21, 23, 26

Information handling: Statistics and probability skills

Probability scale pp233–234

1

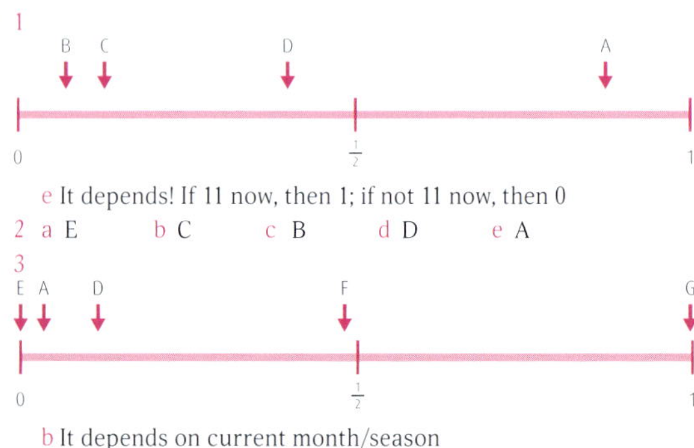

e It depends! If 11 now, then 1; if not 11 now, then 0

2 a E   b C   c B   d D   e A

3

b It depends on current month/season

c as b

4

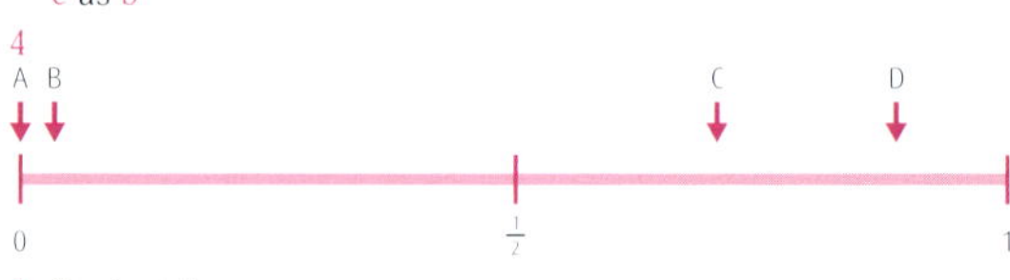

5 Students' own answers

Probability of an event happening pp235–236

	a	b	c	d	e
1	$\frac{1}{6}$	$\frac{1}{2}$	$\frac{1}{3}$	$\frac{2}{3}$	0
2	$\frac{1}{2}$	$\frac{1}{4}$	$\frac{5}{12}$	$\frac{1}{2}$	$\frac{1}{4}$
3	$\frac{1}{2}$	$\frac{3}{10}$	0	$\frac{2}{5}$	$\frac{3}{10}$
4	$\frac{1}{20}$	$\frac{1}{2}$	$\frac{1}{2}$	$\frac{13}{20}$	$\frac{1}{10}$
5	$\frac{1}{7}$	$\frac{2}{7}$	$\frac{3}{7}$	$\frac{4}{7}$	0
6	$\frac{6}{11}$	$\frac{5}{11}$	0		
7	$\frac{7}{20}$	$\frac{2}{5}$	$\frac{1}{4}$	$\frac{3}{5}$	$\frac{3}{4}$
8	$\frac{2}{11}$	$\frac{4}{11}$	$\frac{2}{11}$	$\frac{7}{11}$	0
9	$\frac{1}{6}$	$\frac{1}{4}$	$\frac{1}{4}$	$\frac{2}{3}$	0
10	$\frac{1}{4}$	$\frac{1}{13}$	$\frac{3}{13}$	$\frac{1}{2}$	$\frac{1}{52}$

Relative and expected frequency of an event happening p238

1 50   2 30

3 a 20   b 120   c 60   4 a 12   b 60   c 48

5

	1	2	3	4	5	6
Relative frequency	1	4	5	10	5	5

Biased. Would expect about 5 of each number – but results are '4 heavy'.

6

A	B	C
15	8	7

Biased – lot more As than would be expected.
With the same relative frequency, expect

a 60   b 32   c 28

Making decisions based on chance and uncertainty pp239–241

1 a Not. Throw does not depend on previous outcome.
  b Not. It depends on number of boys/girls in school.
  c Not. A fair die would end up on a six about $\frac{1}{6}$ of time.
  d Not. Do not know how many of each type of biscuit.
  e Not. It depends on the time of year.

2 a Incorrect (even chance)
  b Incorrect (same chance as before $\frac{1}{6}$)
  c Incorrect (he may fall off)
  d Correct

3 No. Each draw is only about 5% and does not depend on previous draws.

4 Pretty safe.
  Choose a safe place to cross
  Use a pedestrian crossing
  Use 'green cross code' or similar
  Look and listen as crossing

5 No. 18 holes per week is about 1000 holes in one year. So with a chance of only 1 in 5000, it is unlikely to get a hole in one in one year.

6 Yes. Very small chance. If a properly organised and supervised event, it should be safe.

7 Probably correct. 100 years would be about 1200 lawn 'mowings'. With 1 in 36000 chance of injury, it should be fairly safe.

GLOSSARY

Acute: an acute angle is less than 90°.

Alternate angles: angles which occur when a straight line intersects parallel lines. They are equal and form a z-shape.

Bearing: indicates which direction to move from a given point, for example "the ship sailed on a bearing of 147°".

Cartesian diagram: a coordinate grid named after Renée Descartes.

Circumference: the distance around the outside of a circle, found using the formula $c = \pi d$.

Class intervals: when arranging widely spread data into a frequency table, class intervals such as 10 – 19, 20 – 29, 30 – 39 and so on make the task simpler.

Commission: a way of rewarding salespersons by giving them a percentage of their sales as part of their salary.

Complementary angles: complementary angles add up to 90°.

Composite: a composite shape is a shape made up of two or more shapes, for example a rectangle and a triangle.

Compound bar graph: an extension of an ordinary bar graph which compares two or more quantities at the same time.

Continuous: continuous data is measured data which can take any value in a given range, for example, the heights of a group of students.

Correlation: correlation describes the connection between two sets of data, for example height and weight.

Corresponding angles: angles which occur when a straight line intersects parallel lines. They are equal and form an f-shape.

Denominator: the bottom number in a fraction, for example 5 in $\frac{3}{5}$.

Direct proportion: two quantities are said to be in direct proportion if, as one quantity increases or decreases, the other increases or decreases at the same rate.

Discount: a discount reduces the price of an item in a sale.

Discrete: discrete data is data which can only take certain values in a given range, for example, the number of goals scored in a football match.

Equation: an equation is a sentence with the verb 'is equal to' in it.

Equilateral triangle: a triangle in which all three sides are equal in length. Also all three internal angles are 60°.

Expression: an expression is a term such as $2x$ or a collection of terms such as $2x + 3y$.

Formula: A rule connecting two or more quantities.

Gradient: can be found using the formula gradient = vertical height /horizontal distance and tells you the slope of a line.

Highest common factor (hcf): the hcf of two or more numbers is the largest factor common to these numbers, for example the hcf of 20 and 30 is 10.

Hire purchase: a way of buying items by paying a small part of the cost, called the deposit, followed by monthly instalments.

Hypotenuse: the longest side in a right-angled triangle. It is opposite the right angle.

Inequation: an inequation is a sentence containing 'is greater than' or 'is less than' in it.

Integer: the set of integers {...–3, –2, –1, 0, 1, 2, 3...} is the set of positive and negative whole numbers and zero.

Interest: interest is paid by banks to its customers as a percentage of the amount they have deposited in an account.

Interest (2): amount charged on, for example, a loan.

Isosceles triangle: a triangle with two equal sides and angles.

Like terms: terms such as $5a$ and $2a$ are like terms and $5a + 2a$ can be simplified leading to $7a$.

Line symmetry: where one half of a shape is the reflection of the other half in an axis of symmetry.

Lowest common multiple (lcm): it is the smallest number which two (or more) numbers divide exactly. For example, the lcm of 4 and 6 is 12.

Mean: an average of a set of data defined by total of all values /number of values.

Median: the middle value in a set of ordered values.

Mode: the most frequent value in a data set.

Net: the net of a solid shape is a flat two-dimensional shape which can be cut out and folded up to make the solid shape.

Numerator: the top number in a fraction, for example 3 in $\frac{3}{5}$.

Obtuse: an obtuse angle is greater than 90° and less than 180°.

Perpendicular: two lines are perpendicular if they are at right angles to one another.

Polygon: A many-sided figure. For example, a hexagon (6 sides).

Power: to raise a number to a power is to multiply it by itself a number of times. For example, $3^4 = 3 \times 3 \times 3 \times 3$

Prime number: a number which has exactly one pair of factors.

Prism: a prism is a solid shape which has opposite ends that are parallel and congruent, for example a cuboid.

Probability: a measure of how likely an event is to happen.

Proportion: a part or a share of a whole. For example, there are 3 green marbles and 2 red marbles in a bag. The proportion of green marbles is $\frac{3}{5}$.

Quadrilateral: any four-sided figure such as a square, kite, rhombus, etc.

Range: for any numerical data set, the range = highest value – lowest value.

Ratio: shows the relationship between two (or more) amounts. For example £100 is shared in the ratio 2:3

Reflex: a reflex angle is greater than 180° and less than 360°.

Right angle: an angle which is exactly 90°.

Rotational symmetry: a shape has rotational symmetry when it fits its outline as it turns (or rotates).

Scale factor: the size of an enlargement (or reduction) is described by its scale factor. For example, a scale factor of 2 means that all the lengths in the new shape are twice the lengths of those in the original shape.

Scattergraph: a scattergraph is a statistical diagram which can be used to compare two sets of data by plotting a set of points on a coordinate grid.

Sequence: A list of, for example, numbers which follow a given rule.

Similar: two shapes are said to be similar when one shape is an enlargement or reduction of the other.

Square root: a number which, when multiplied by itself gives the original number, for example $\sqrt{16} = 4$ (or -4)

Stem and leaf diagram: an ordered statistical diagram in which each member of the data set is split into a 'stem' and a 'leaf'.

Straight angle: an angle which is exactly 180°.

Supplementary angles: supplementary angles add up to 180°.

Tangent: a tangent to a circle is a straight line which touches the circle at one point only. This point is called the point of contact.

Theorem of Pythagoras: states that in a right-angled triangle, the square on the hypotenuse is equal to the sum of the squares on the other two sides, usually given by the formula $a^2 + b^2 = c^2$.

Transformation: a movement of a shape, for example translation, enlargement or reduction.

Trend: the trend of a graph indicates the general direction that the graph is going, for example increasing or decreasing.

Value added tax: value added tax, or vat for short, is added to the cost of certain goods.

Vertically opposite angles: vertically opposite angles are equal. They are formed when two straight lines intersect.

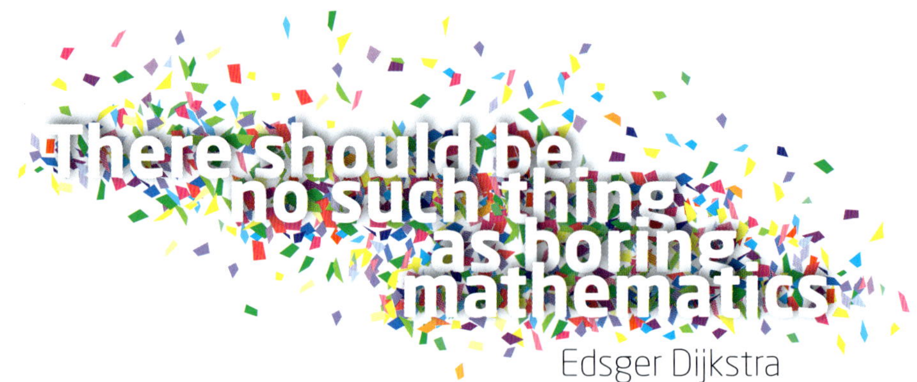